Irradiation Effects
in
Nuclear Fuels

American Nuclear Society
and
U.S. Atomic Energy Commission
(Division of Technical Information)

MONOGRAPH SERIES ON
NUCLEAR SCIENCE AND TECHNOLOGY

John Graham
Series Editor American Nuclear Society

ADVISORY COMMITTEE

MONOGRAPH TITLES AND AUTHORS

IRRADIATION
EFFECTS
IN
NUCLEAR FUELS

An AEC Monograph

J. A. L. ROBERTSON

Atomic Energy of Canada, Ltd.

Prepared under the direction of the
American Nuclear Society
for
The Division of Technical Information
United States Atomic Energy Commission

GORDON AND BREACH, SCIENCE PUBLISHERS
New York / London / Paris

Foreword

This monograph is one of a series developed through the joint efforts of the American Nuclear Society and the Division of Technical Information of the U.S. Atomic Energy Commission. The purpose of the undertaking is to cover very specific areas of nuclear science and technology and thus help to advance the peaceful applications of nuclear energy.

While the monographs are primarily directed toward the operational scientist or engineer concerned with the applications of nuclear energy, they should also be helpful to students of science and engineering who otherwise might have little opportunity to study information within the special area of each monograph.

In looking forward to many dramatic accomplishments in peaceful uses of nuclear energy, the American Nuclear Society is pleased to cooperate with the U.S. Atomic Energy Commission in developing this series of monographs to help reach these achievements.

John Graham
Senior Staff Editor
American Nuclear Society

Preface

The purpose of this monograph is to assemble in one source facts and theories relating to the irradiation of nuclear fuels. I thereby hope to show that a relatively few phenomena account for most of the observed behavior of vastly different types of fuel. That is not to say that the individual phenomena are simple or even well understood. Indeed, the large number of potentially important variables in any instance, and the experimental difficulties in controlling them, have probably discouraged many appropriately qualified persons from investigating these interesting phenomena.

To keep within manageable proportions, I have excluded liquid and gaseous fuels. Even with solid fuels, I have had to confine to a minimum the consideration of the physical properties and fabrication methods of the unirradiated materials. Similarly, I have not attempted to recognize all the irradiations that have been reported, but only those that have some general application or that help us understand the phenomena. Admittedly, this selection has resulted in extra weight being given to those observations that can be interpreted. Where I have ignored results (or authors' conclusions) it is usually because not enough was known of the experimental conditions to determine whether the results were anomalous or not.

For almost any particular fuel treated here, there are already in existence one or more excellent reviews that often describe the irradiations in greater detail, besides providing information on related topics: Many of these are given as references. Some additional sources of information on nuclear fuels are listed at the end of the references.

To help illustrate the text, I have used "thumbnail"

sketches in lieu of formal figures. Some are intended to make immediately obvious a simple concept that would appear complicated if one relied on words alone. Others summarize experimental results, where it is the functional relation between the variables, and not their actual values, that is important. Conventional graphs are still used where the values are important. I am indebted to J. H. Moffatt for the preparation of both types of drawing.

In preparing this monograph I was encouraged by the opinion of the Scientific Advisory Committee to the President of the United States that critical reviews of both report literature and the related open literature "play an important part in easing the information crisis. They serve the special needs of both the established workers in a field and the graduate student entering the field, as well as the general needs of the nonspecialist." The same committee, however, sounded a warning: "A simple but urgent suggestion to authors is to refrain from unnecessary publication. The literature has been and always will be cluttered with poor and redundant articles."

For whatever may be of value in this monograph, I wish to acknowledge the benefit I have received over the years from many discussions with my colleagues both here and elsewhere, especially A. S. Bain, B. G. Childs, J. R. MacEwan, M. J. F. Notley, A. M. Ross, and O. J. C. Runnalls. I am also most grateful to the referees appointed by the American Nuclear Society (T. K. Bierlein, M. L. Bleiberg, J. E. Gates, J. H. Kittel, and J. B. Sayers) all of whom provided truly helpful comments and constructive criticism. I appreciate the congenial and stimulating atmosphere at the Chalk River Nuclear Laboratories, largely attributable to the direction of Dr. W. B. Lewis. Finally I thank my family for their understanding during the last two years.

Deep River, Ontario
May, 1965

Contents

1

Introductory Survey

1.1 Requirements of a Nuclear Fuel

Any material containing fissile atoms can serve as a nuclear fuel. In naturally occurring uranium 1 out of every 139 atoms is the isotope ^{235}U which, on capturing a neutron, may subdivide releasing a large amount of energy. The ^{235}U content can be artificially increased in enriched uranium, but at a cost representing the energy needed for isotopic separation. The bulk of natural uranium is the isotope ^{238}U which will fission only if the neutron it captures has sufficient energy, i.e., it is fissionable rather than fissile. There is, therefore, a distinction between "fast fission" in a flux of sufficiently energetic (or fast-moving) neutrons, and "thermal fission" in a flux of neutrons that are approximately in thermal equilibrium with their environment.

Although capture of a thermal neutron does not cause fission in an atom of ^{238}U, the product becomes a plutonium isotope ^{239}Pu, which can undergo thermal fission. Similarly, ^{232}Th, the main constituent of naturally occurring thorium, is the source of another fissile uranium isotope ^{233}U. Thus, ^{238}U is termed a "fertile material", in addition to being a fissionable material. Continued capture of neutrons by either fissile or fissionable materials can produce heavier transuranic elements, some isotopes of which are fissile. The presence of these other fissile materials can affect the neutron balance of a reactor, but ^{235}U, ^{239}Pu, and ^{233}U are the only ones normally present in sufficient concentration to be important to the fuel's behaviour in other respects.

1

Reactor physics determines the best fissionable material for a given reactor type, the amount (if any) of fertile material that should be incorporated, and the extent to which the two should be segregated. As a very rough generalization, for application to central power stations the fissile-to-fertile atomic ratio is commonly around $1:50$ in thermal reactors and $1:5$ in fast ones. For mobile power reactors, or for experimental and testing reactors, the fissile content for each type is generally higher so that, in the limit, some fuels incorporate no fertile material. If little fertile material is intimately associated with the fissile material, the energy release per unit volume could be very high. For this reason, and because the presence of fertile material is usually desirable, fuels entirely composed of fissile material are rarely required. To avoid heat fluxes that are too high for the coolant's capabilities, or excessive temperatures in the fuel, it is sometimes advisable to disperse the fissile atoms in a diluent material.

In thermal or intermediate (between thermal and fast) reactors, a moderator is needed to slow down the neutrons produced by fission to thermal energies where they are most efficient in promoting further fission. In almost all reactors, sheathing material is required to protect the fuel from corrosion, to contain highly active fission products, and to maintain the fuel's configuration. Some dispersion fuels, therefore, make the moderator or sheathing material serve the additional function of diluting the fissile material, but in others any diluent must be judged entirely on its own merits.

An important criterion for judging any material put into the fuel that does not fulfil an essential nuclear function is its capture cross section for neutrons. Generally, the cross section for fast neutrons is much smaller than that for thermal neutrons [the latter is expressed in barns (b), i.e., 10^{-24} cm^2, and the former in millibarns (mb)]. Since the capture cross section of an isotope depends on the neutron energy in a manner characteristic of the isotope, it is quite possible that an element with a relatively high thermal cross section may have a relatively low fast one,

and vice versa. In fast reactors the cross section for
scattering neutrons is also important.

Designers are often willing to tolerate higher concentrations of parasitic materials in fast reactors than in thermal
ones, because of their lower capture cross sections relative
to that of the fuel. However, in any system, a neutron
captured without causing fission, even eventually, is wasted
and has to be replaced by provision of fresh fissionable
material. This fact is emphasized by calculating a cost
for neutrons, either individually or by weight. The cost
depends on several factors, including the fissionable material being used and the nature of the fuel cycle but, in
illustration, a figure of \$2860/g can be quoted (Lewis,
1961). For comparison, 1 kg of stainless steel in a thermal
flux of 6×10^{13} n/(cm^{2}sec) captures 0.1 g of neutrons
per year.

The fuel density can be another important criterion. In
some reactors a low density fuel would require more
sheathing material and more containment material for
the coolant, both of which capture neutrons to some
extent. In a fast reactor any light elements present in a
low density fuel would moderate the fast neutrons, thereby
decreasing the efficiency of conversion of fertile to fissile
material. Thus, high density fuels are usually desirable,
but there are exceptions to this generalization. For instance, increasing the volume of the fuel at the expense
of a coolant with a relatively high capture cross section
can be beneficial.

Ignoring the exceptions helps in classifying, for consideration in subsequent sections, the great variety of fuels
that have been investigated. Since uranium metal is ideal
from the standpoint of reactor physics, having no parasitic
additions and the maximum density, it will be treated
first. Even minor amounts of alloying elements have to
be justified by their improving the material properties
in some way. Where an addition to the uranium is necessary
to enable the fuel to achieve its desired performance, it is
often preferable to sacrifice density rather than accept
significant amounts of parasitic material: This argument

has led to the widespread use of uranium compounds, such as the oxide and carbide, as nuclear fuels. Finally, consideration will be given to dispersion fuels, most of which have both a low uranium density and an appreciable parasitic content. Fissionable oxides and carbides diluted with inert oxides and carbides, respectively, might logically be regarded as dispersion fuels, but, for convenience, they will be grouped with the appropriate fissionable compounds.

The choice of a fuel material for a particular application depends on other factors; the cost of the constituents, the ease of fabrication and reprocessing, as well as the dimensional stability, corrosion resistance, and strength at temperature and under irradiation. For power reactors, neither low fuel costs, good neutron economy, nor good operating performance alone is sufficient to assure selection, but some optimum combination of all three is required to minimize the cost of electricity supplied. For other reactors the criterion is less easily defined, but a similar compromise is usually necessary. The purpose of most irradiation experiments on fuel is to provide information that permits a rational assessment of the competing factors.

1.2 Mechanisms of Irradiation Damage

Certain effects of irradiation are common to all solids. Several of the phenomena have already been thoroughly studied in the irradiation of nonfissionable material, and the isotopic distribution of fission products has been well established by radiation chemistry. In fissionable solids three major sources of irradiation damage can be distinguished: displacements of atoms and electrons from normal lattice sites; the release of a large amount of heat in a small volume; and the presence of fission-product atoms acting as impurities in the original material.

When a heavy atom fissions, about 200 MeV of energy is released, distributed approximately as follows:

	MeV
Kinetic energy of fission-product atoms	169
Kinetic energy of fission neutrons	5
Instantaneous gamma radiation	5
Fission-product decay energy	12
Beta and gamma radiation following neutron capture	8
	199

Of these components, the first three occur at the time of fissioning while the last two occur later. In addition, neutrons carry about 11 MeV out of the reactor. Most of the energy is thus carried by the fission products, which have atomic weights from 75 to 160. In passing through the crystal lattice of the fuel, these fission products, which are positive ions for most of the time they are moving, lose much of their energy by interaction with electrons in the fuel. The consequent ionization of the lattice atoms is relatively unimportant: Even in nonmetallic fuels the electrons are usually sufficiently mobile for rapid recovery of the electron distribution (Sec. 3.10).

Much more important to the fuel's behaviour are collisions of the fission products with fuel atoms, knocking the latter off their lattice sites. The resulting "primary knock-ons" can acquire sufficient energy in the collision for them, in turn, to eject further atoms, termed "secondary knock-ons". The process terminates when an ejected atom has insufficient energy to displace another atom from its lattice site. Since this critical energy is probably under 100 eV for most fuels, each fission event may produce as many as 10^6 knock-ons. The fast neutrons released during fission also cause atomic displacements, but they are far less effective by virtue of the smaller amount of energy they carry.

An atom ejected from its lattice site leaves a vacancy: On coming to rest, the atom may occupy a vacancy left by another knock-on or it may lodge in an interstitial site (Fig. 1–1). In alloys or compounds there can be both vacancies and interstitials of each element present. In addition, atoms of one element can be displaced into lattice sites properly occupied by an atom of another element, thereby disordering the original structure. These individual imperfections in the crystal lattice are termed "point defects".

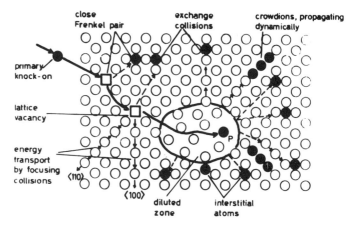

Fig. 1–1. Schematic picture illustrating the terminology of damage mechanisms. A Frenkel pair consists of a vacancy and an interstitial atom: An exchange collision occurs when a lattice atom is knocked into an adjacent interstitial position by another atom, which then occupies the vacant site: Consecutive exchange collisions along a line of atoms produce a crowdion. (After Seeger (1962), courtesy International Atomic Energy Agency.)

When a knock-on is impelled along a crystallographic direction on which no lattice sites are located, it can travel a long distance through the lattice. This phenomenon is known as "chanelling". Conversely, a knock-on that moves along a direction for closest approach of atoms in the lattice is brought to rest immediately, but the displacement energy continues to be transmitted along the line of atoms. Even if the movement of the initial knock-on does not quite lie along the favoured direction, each successive collision tends to bring the trajectory of the next knock-on nearer to the direction of closest approach. This "focussing" has been observed experimentally with several materials, including UO_2 (Nelson, 1963). These collision mechanisms are illustrated diagramatically in Fig. 1-1. The net result of both chanelling and focussing is to produce a widely separated vacancy/interstitial pair. Thus, the core of the damaged region, or "fission spike", is rich in vacancies, while the surrounding shell contains a relatively high concentration of interstitials.

Undoubtedly many of the interstitials and vacancies annihilate each other, even while further atoms are being displaced. Also, point defects of the same type can agglomerate to form "interstitial clusters" or "vacancy clusters". If the cluster forms a sheet one atom thick, the defect can equally well be described as a small dislocation loop; otherwise, the cluster may form a stacking fault tetrahedron or a minute pore. In an ionic compound the occurrence of dislocation loops is confined to certain crystallographic planes if adjacent planes of atoms of the same charge are to be avoided. The tendency for cluster formation is probably helped by the segregating effect of chanelling and focussing.

Some of the kinetic energy of a fission fragment is thus converted into potential energy of the various defects, but most appears as thermal energy of the atoms in the fission spikes along the track of the fragment. The extent of a spike in space and time depends on arbitrary definitions of what constitutes its limit, but a typical one can conveniently be thought of as being 10^4 atoms long, between 10 and 100 atoms in diameter, and lasting 10^{-11} sec. In this volume the lattice is thoroughly disrupted and the large thermal vibrations of the atoms cause considerable expansion. Consequently, the spike is in compression while the surrounding matrix is subject to tensile stresses. The thermal diffusivity of the fuel determines the time taken for the heat in the spike to be dissipated uniformly by lattice vibrations.

The range of a fission-product atom decreases with increasing density of the material in which it travels, but a value of eight microns is reasonably representative. Another factor affecting the range is the mass of the fission-

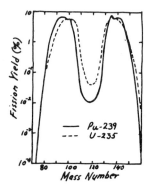

product atom itself. From each fission event there are emitted two product atoms of unequal mass but having roughly equal momentum. The lighter one starts with the greater kinetic energy and thus penetrates further. The frequency of occurrence as a function of mass number is sketched alongside. For most purposes the only masses produced in sufficient quantity to be effective are those close to either of the maxima, where the atomic yield is one to ten percent of the number of fissions (not the number of fission-product atoms, which is twice as many). Since a given element is associated with only a few mass numbers, even a rough mass distribution of the type sketched above shows which elements are relatively abundant. These are, in order of increasing atomic number, from bromine to cadmium and from antimony to gadolinium.

Because an atom of a particular isotope can be lost, or formed, by either neutron capture or radioactive decay, the amount of each fission product depends on the neutron flux during irradiation and the time since fission occurred. Thus, the precise calculation of the amount of any given element is a complicated one, but, fortunately, the results have been tabulated by various authors, e.g., Blomeke and Todd (1958); Katcoff (1960); and Bromley (1963). The sketch indicates that the fission-product yield distribution is characteristic of the fissionable material, with ^{235}U, ^{239}Pu, ^{233}U, and every other fissionable isotope having its own curve. The distribution also depends on the energy of the neutron that causes fission. The total energy release per fission varies slightly between the fissionable isotopes, but it is near 200 MeV for them all.

1.3 Experimental Measurements

Fuel irradiations are notorious for the irreproducibility of their results. This situation is partly due to the large number of potential variables, several of which are most difficult to control adequately. Knowledge of two particular variables, and how they vary with time, is of the greatest importance: the burnup, or exposure; and the temperature distribution of the fuel. The number of fissions that have occurred per unit volume of fuel determines the damage experienced by the crystal lattice, as well as the concentration of fission products. The fission density per unit time determines the power generation of the fuel and, hence, affects its temperature distribution.

Indeed, the interdependence of temperature and rate of burnup can result in serious experimental difficulties in fuel irradiations. For instance, an error of only ten percent in the burnup rate would cause an error of about 100°C at 1500°C in irradiating UO_2. For a process with an activation energy of 100 kcal/mole, a normal value for diffusion-controlled phenomena in UO_2, such an error in temperature would produce an error of an order of magnitude in the process under study. By comparison, errors of a factor of 2 in exposure are usually of little consequence in irradiations of nonfissionable materials.

When the importance of knowing the exposure accurately is realized, it seems nearly incredible that four completely different units are in common use to describe this quantity: integrated neutron flux; percentage burnup; fission density; and energy density. Even worse, each unit can have different values depending on how it is defined, yet many experimenters do not state the convention they employ.

Measurements of the neutron flux in the reactor, either by continuously reading instruments or by activation monitors, can yield the integrated neutron flux at the specimen's position for the duration of its irradiation. Since the specimen's presence perturbs the flux being measured, corrections have to be calculated and applied to give the flux experienced by the specimen. The integrated flux is commonly expressed as the "*nvt*", in neutrons/cm^2 (*n* is the density of neutrons, *v* their mean velocity, and *t* the time of exposure) even though *n* and *v* may never be determined separately. Where a single specimen, or closely similar specimens, are exposed for different periods or in different fluxes, the integrated flux can provide a useful means of comparison. However, specimens of different composition or of different fissile enrichment exposed to the same integrated flux would experience different degrees of damage. Also, the spectrum of neutron energies can vary appreciably between reactors, so that the same integrated neutron flux would not always produce the same number of fissions in similar specimens.

To compare the relative amounts of damage in different fuel materials, a better unit is the percentage of all atoms (fissionable, fertile, and diluent) that have been fissioned, termed the burnup. However, for other purposes, it is more important to know the percentage of fissile atoms or of all fissile plus fertile atoms that have been fissioned, and the same term, burnup, is used. Burnup may even refer to the percentage of fissile atoms that have captured a neutron, whether to cause fission or not. Utter confusion results when a burnup is quoted without precise definition of its meaning. Analyses of the irradiated fuel for fission products or by mass spectrometry to determine the isotopic ratios in the fissionable and fertile materials provide a direct measure of the burnup, but certain constants, such as fission yields, must be known. When experimental errors are fully assessed, these methods do not necessarily give more accurate values for the burnup than less direct ones.

To relate the burnup and the integrated flux, consider N atoms of a fissile isotope whose cross section for neutron capture is σ_c and for fission σ_f. The rate of loss of the original fissile atoms is

$$\frac{dN}{dt} = -Nnv\sigma_a, \qquad \sigma_a = \sigma_c + \sigma_f, \qquad (1\text{–}1)$$

where σ_a is the neutron absorption cross section, so that by integrating the percentage loss is obtained as $100[1 - \exp(nvt\sigma_a)]$. But only a fraction of these has been lost by fission and therefore the burnup, expressed as the percentage of original fissile atoms that has been fissioned, is $100\,(\sigma_f/\sigma_a)[1 - \exp(-nvt\sigma_a)]$. Where $nvt\sigma_a$ is much less than unity, an approximation for the exponential is valid.

Thus, the same burnup becomes 100 $nvt\sigma_f$. For ^{235}U in most thermal fluxes, σ_f approximately equals 550 b, so the burnup is approximately $5.5 \times 10^{-20} \times (nvt)\%$ of the ^{235}U.

If the fissile atoms are uranium isotopes,

$$100\ (N_f N_u)\ nvt\ \sigma_f$$

expresses the burnup as a percentage of all the uranium atoms originally present, where N_f and N_u are the numbers of atoms of the fissile isotopes and all uranium isotopes, respectively, in a given volume of the specimen. For natural uranium, N_f/N_u is 1/139, and thus this burnup is approximately $4 \times 10^{-22} \times (nvt)\%$. Expressed as a percentage of all atoms in the fuel, the burnup is $100\ (N_f/N_t)\ nvt\ \sigma_f$, where N_t is the total number of all atoms present in the same volume. Thus, any of the definitions of burnup may be derived from the integrated flux and a knowledge of the fission cross section. The derivation here has been greatly simplified, but the result is generally applicable if a consistent convention is used for measuring both the flux and the cross section, for instance, that proposed by Westcott (1960).

Another unit, the fission density in fissions/cm^3, has gained increasing favour. It is just as useful as the burnup of all atoms in describing the damage, yet is less open to misinterpretation. Some opportunity for ambiguity occurs in dispersion fuels. Particularly with coated particles, the volume of fuel considered may be only the fissionable phase or may include the coating and matrix. Normally, the fission density is the number of fissions divided by the total volume of fuel plus any diluent, but excluding sheathing. The fission density is the product of the *fraction* of uranium atoms fissioned and the density of uranium atoms in the fuel. Thus, the fission density equals $(N_f/N_u)\,nvt\sigma_f(\rho N_A/M')$, where ρ is the fuel density, M' the molecular weight of the fuel divided by the number of uranium atoms in the molecular unit, by N_A Avogadro's Number (6.02×10^{23}/mole). From the earlier conversion the fission density can also be related to the burnup as $(\rho N_A / 100 M')$ ($\%$ U fissioned).

Finally, the burnup or exposure is frequently given as an energy density in units of MWd/t. Unfortunately, different tons (or tonnes) are in common use and the mass considered may be that of the total fuel or its fissionable content. The temperature distribution in the fuel depends on the heat released within it, yet this can differ significantly from the heat generated by fission since as much as ten percent of the energy of fission may be carried out of the fuel by neutrons and gamma radiation (Sec. 1.2). To compare damage in fuels, the most reasonable version is probably megawatt days (MWd) released in the fuel per metric tonne (10^3 kg) of total fuel, but megawatt days total generation per metric tonne of contained uranium is perhaps more often found. Once again, the minimum requirement is that the unit used be clearly stated. The most direct method of obtaining the energy density is by measuring the power output of the fuel calorimetrically and integrating with respect to time. However, this method of evaluating the exposure also requires the calculation of correction factors in practice and is susceptible to experimental errors.

A single fission generates a total of approximately 200 MeV (Sec. 1.2), which equals 3.7×10^{-16} Wd. Since MWd/tonne is the same as Wd/g,

MWd generated/tonne fuel

$$= 3.7 \times 10^{-16} \frac{1}{\rho} \times (\text{fissions/cm}^3), \quad (1\text{--}2)$$

and

MWd generated/tonne U

$$= 3.7 \times 10^{-16} \frac{M'}{\rho A} \times (\text{fissions/cm}^3), \quad (1\text{--}3)$$

where A is the atomic weight of uranium. Almost all the kinetic energy of the fission fragments and the beta particles is released within the fuel, but many of the fission neutrons and gamma rays escape from the fuel to release their energy in the coolant, moderator, and other reactor components. Thus, the heat per fission that is released within the fuel, E_f MeV/fission, varies with the geometry and properties of the fuel and its surroundings, but in a typical thermal reactor the value is 180 MeV/fission. The burnup, in terms of heat released in the fuel, is then

MWd within fuel/tonne fuel

$$= 1.85 \times 10^{-18} \frac{E_f}{\rho} \times (\text{fissions/cm}^3), \qquad (1\text{--}4)$$

and

MWd within fuel/tonne U

$$= 1.85 \times 10^{-18} \frac{M'E_f}{\rho A} \times (\text{fissions/cm}^3). \quad (1\text{--}5)$$

Early in the derivations an approximation eliminated the exponential function. However, the exponential could easily have been retained throughout if more exact relations were required, and only those conversions involving the integrated flux are affected by the approximation. Ignoring fission of fresh fissile atoms produced by the capture of neutrons in fertile atoms is an oversimplification that is less easily corrected. The error is more important in fuels of low enrichment and at high burnup. As a rough guide, overlooking the plutonium buildup in natural uranium can result in the fission density being underestimated by ten percent by an exposure of 0.1 % U burnup by fission or 1000 MWd within fuel/tonne U. The fission density is also underestimated by the amount of fast fission, which normally contributes a few percent to the total in thermal reactors. In fast reactors, the analogous calculations would use the cross sections for fast neutrons.

Universal acceptance of a single unit for burnup is unlikely. Reactor designers are concerned more with the power generated by the fuel than with its fission density:

Table I-I

% U atoms fissioned	$= C_1 \times (n/cm^2)$ *	;	$C_1 =$
% all atoms fissioned	$= C_2 \times (\%U \text{ atoms fissioned})$	;	$C_2 =$
	$= C_3 \times (n/cm^2)$ *	;	$C_3 =$
Fissions/cm³	$= C_4 \times (\% \text{ all atoms fissioned})$	;	$C_4 =$
	$= C_5 \times (\% \text{ U atoms fissioned})$	;	$C_5 =$
	$= C_6 \times (n/cm^2)$ *	;	$C_6 =$
MWd within fuel/tonne U	$= C_7 \times (\text{fissions/cm}^3)$	;	$C_7 =$
	$= C_8 \times (\% \text{ all atoms fissioned})$	;	$C_8 =$
	$= C_9 \times (\% \text{ U atoms fissioned})$	;	$C_9 =$
	$= C_{10} \times (n/cm^2)$ *	;	$C_{10} =$
MWd within fuel/tonne fuel [†]	$= C_{11} \times (\text{MWd within fuel/tonne U})$;		$C_{11} =$

* Involves linear approximation for exponential; examples assume $\sigma_f = 550$ b for ^{235}U.

$N_f =$ Number of fissile atoms/volume fuel
$N_u =$ Number of uranium atoms/volume fuel
$N_t =$ Total number of atoms/volume fuel

Burnup Conversion Factors

	Uranium	$UO_2 \#$	UC
$00 \dfrac{N_f}{N_u}\sigma_f$	$5.5 \times 10^{-20}\dfrac{N_f}{N_u}$	$5.5 \times 10^{-20}\dfrac{N_f}{N_u}$	$5.5 \times 10^{-20}\dfrac{N_f}{N_u}$
$\dfrac{N_u}{N_t}$	1	1/3	1/2
$00 \dfrac{N_f}{N_t}\sigma_f$	$5.5 \times 10^{-20}\dfrac{N_f}{N_t}$	$5.5 \times 10^{-20}\dfrac{N_f}{N_t}$	$5.5 \times 10^{-20}\dfrac{N_f}{N_t}$
$.02 \times 10^{21}\dfrac{\rho}{M'}\dfrac{N_t}{N_u}$	4.8×10^{20}	6.9×10^{20}	6.6×10^{20}
$.02 \times 10^{21}\dfrac{\rho}{M'}$	4.8×10^{20}	2.3×10^{20}	3.3×10^{20}
$.02 \times 10^{23}\dfrac{\rho}{M'}\dfrac{N_f}{N_u}\sigma_f$	$26.4\dfrac{N_f}{N_u}$	$12.6\dfrac{N_f}{N_u}$	$18.2\dfrac{N_f}{N_u}$
$.85 \times 10^{-18}\dfrac{M'E_f}{\rho A}$	$10^{-19}E_f$	$2 \times 10^{-19}E_f$	1.4×10^{-19}
$.1 \times 10^4 \dfrac{E_f}{A}\dfrac{N_f}{N_u}$	$46.5\,E_f$	$139\,E_f$	$93\,E_f$
$.1 \times 10^4 \dfrac{E_f}{A}$	$46.5\,E_f$	$46.5\,E_f$	$46.5\,E_f$
$.1 \times 10^6 \dfrac{E_f}{A}\dfrac{N_f}{N_u}\sigma_f$	$2.5 \times 10^{-18}E_f\dfrac{N_f}{N_u}$	$2.5 \times 10^{-18}E_f\dfrac{N_f}{N_u}$	$2.5 \times 10^{-18}E_f\dfrac{N_f}{N_u}$
$\dfrac{A}{M'}$	1	0.88	0.96

$\#$ 95% of theoretical density
$\dagger$ For MWd generated, $E_f = 200$ MeV/fission, approximately

ϱ = Fuel density
σ_f = Fission cross section of fissile atoms
E_f = MeV/fission, released in fuel

A Atomic weight of uranium
M' Molecular weight of fuel divided by number of uranium atoms per molecular unit

Reactor physicists are concerned more with the percentage of fissile atoms lost by all processes than with the percentage fissioned. Even to those primarily interested in the behaviour of the fuel, the percentage of uranium atoms fissioned sometimes conveys more meaning than the fission density. For instance, 70% burnup of the uranium atoms in a UO_2-steel cermet has a significance not immediately apparent in 3×10^{21} fissions/cm^3. In the following sections, the units adopted represent a compromise between current usage and sufficient uniformity to permit intercomparison. The conversion factors between the various burnup units are collected in Table I-1. Where convenient, simple equivalents are provided in the text.

The power density h in the fuel averaged over the duration of the irradiation can be calculated from the burnup in any of its forms. Thus,

$$h = \rho \times (\text{MWd within fuel/tonne fuel})/(\text{days irradiation}), \quad (1\text{--}6)$$

or

$$h = 1.85 \times 10^{-18} E_f \times (\text{fissions/cm}^3)/(\text{days irradiation}), \quad (1\text{--}7)$$

or

$$h = 1.11 \times 10^4 \frac{E_f \rho}{M'} \times (\%\,\text{U fissioned})/(\text{days irradiation}), \quad (1\text{--}8)$$

or

$$h = 9.65 \times 10^{10} E_f \sigma_f \frac{N_f}{N_u} \frac{\rho}{M'} \times (nv). \quad (1\text{--}9)$$

In a more rigorous derivation, allowance should be made for the contributions from fertile material and fast fission. Also, gamma radiation from surrounding fuel will generate power in the specimen: A typical value for gamma heating is 1 W/g, but variation by an order of magnitude in either direction is possible. Naturally, these relations only provide the mean power density through the volume of fuel considered. In large specimens, or ones that are highly absorbing for neutrons, separate calculations are usually required to obtain the variation in power density through the fuel. Similarly, knowledge of the irradiation history is needed, if the power density at any particular time is to be calculated.

Most textbooks on heat transfer derive the temperature distribution in simple geometries from knowledge of the power density and boundary conditions. Thus, in a cylindrical rod (the commonest form for a fuel element) of constant thermal conductivity λ, the temperature θ is a parabolic function of the radial coordinate r, i.e.,

$$\theta_r = \theta_s + \frac{h}{4\lambda}(s^2 - r^2), \qquad (1\text{--}10)$$

and

$$\theta_0 = \theta_s + \frac{hs^2}{4\lambda}, \qquad (1\text{--}11)$$

where s is the radius of the rod, θ_s the temperature at the surface, and θ_0 the temperature at the centre. In practice, the power density varies through the fuel as the neutron flux producing it is attenuated by traversing an absorbing medium. However, as long as reactor physics can determine the neutron flux distribution, the temperature distribution can be derived: The mathematics may be more complicated, but no new principle is involved.

Where the thermal conductivity is not constant, but is a known function of temperature, the temperature distribution can still be obtained by using an earlier stage in the derivation of Eq. (1–10), viz.,

$$\int_{\theta_s}^{\theta_r} \lambda d\theta = \frac{1}{4}h(s^2 - r^2), \qquad (1\text{--}12)$$

and

$$\int_{\theta_s}^{\theta_0} \lambda d\theta = \frac{1}{4}hs^2 . \qquad (1\text{--}13)$$

However, the thermal conductivity is usually decreased by irradiation damage, and its value is often unknown in fuel irradiations. Under these conditions, the function

$\int_{\theta_s}^{\theta_0} \lambda d\theta$ can be useful as a measure of the severity of an irradiation. For two specimens of the same fuel irradiated with the same surface temperature the "integrated conductivity" provides a relative measure of their centre temperatures, even if they are of different size and fissile enrichment. It is not necessary to know the conductivity or its temperature dependence, but only to determine the power output of the fuel. Irradiation experiments can associate certain physical observations, e.g., fuel expansion, with values of the integrated conductivity instead of temperature.

   If the power density is constant through the fuel, the power output per unit length of cylinder q can be expressed as

$$q = 2\pi s w = \pi s^2 h, \qquad (1\text{--}14)$$

where w is the heat flux per unit area of the fuel's surface. Therefore,

$$\int_{\theta_s}^{\theta_0} \lambda d\theta = \frac{1}{4} h s^2 = \frac{1}{2} w s = \frac{1}{4\pi} q, \qquad (1\text{--}15)$$

from which it is apparent that the power per unit length is proportional to the integrated conductivity, and so it also provides a measure of the centre temperature, regardless of variations in fuel diameter and power density. For the more general case of variable power density,

$$\int_{\theta_s}^{\theta_0} \lambda d\theta = \frac{1}{4\pi} q F(0), \qquad (1\text{--}16)$$

where $F(0)$ is a factor less than unity that depends on the radius and certain nuclear properties of the fuel (but not the conductivity). The power per unit length is now only approximately proportional to the integrated conductivity, but the approximation is a good one for small diameter rods of low enrichment. A prior stage in the derivation of Eq. (1–16) provides the temperature as a function of the radial coordinate

$$\int_{\theta_s}^{\theta_r} \lambda d\theta = \frac{1}{4\pi} q F(r), \qquad (1-17)$$

where $F(r)$ is now a function of r as well. The derivation of $F(r)$ and the application of the integrated conductivity to simple geometries, with and without flux attenuation, are discussed elsewhere (Robertson, 1961).

Equation (1–16) expresses a most important fact. The maximum power that can be extracted from unit length of a fuel rod is not determined by either the size, the maximum permissible temperature, or the conductivity alone, but rather by the integral of the conductivity with respect to temperature over the range from surface temperature to maximum permissible temperature. Thus, a low conductivity fuel can provide just as much power per unit length as one with high conductivity, if the permissible temperature range is correspondingly greater.

Anyone surveying the available information on fuel irradiations cannot fail to be impressed by the tremendous amount of effort that has been wasted. All too often a fuel irradiation merely tells that a given design is, or is not, adequate under certain ill-defined test conditions. Comparison with other irradiations may be vitiated by lack of knowledge of the specimens' initial conditions. Similarly, the conditions under which the specimens operated may not be well known. Fortunately, experimental techniques are now established for measuring several variables in the reactor, including two of the most important — power output and temperature.

The power output of the fuel can be determined by measuring the flow and temperature rise of coolant passing over the fuel, but the value obtained is an average for all the fuel concerned. The power output at a specific position can be deduced from analysis by a mass spectrometer of the fissionable isotopes, from radiochemical analysis of the fission products, or from activation monitors, but these values are averages over the duration of the irradiation. However, modifications of the activation monitors can provide continuous readings of the instantaneous neutron flux at a selected position. In one, the activity from a stream of argon is measured after the gas has circulated in a tube with an enlarged cross section at the appropriate point (Carroll, 1962). Another contains a wire made from a metal that becomes a beta emitter when activated by neutron capture, e.g., vanadium. A thin metal tube surrounding the wire, but insulated from it, collects the beta particles (Mitelman et al., 1961; Hilborn, 1964). Leads from the wire and the tube permit the resulting current, which depends on the neutron flux, to be measured outside the reactor.

Suitable thermocouples exist for application in most fuels, even for temperatures as high as 2000°C. On occasions, thermocouples have also been used as electrical leads to measure the resistance and thermoelectric power of fissionable specimens during irradiation. Lyons et al. (1964a) developed a gas thermometer for use in UO_2 to near its melting point (2800°C): The gas bulb consisted of a rhenium tube along the axis of a cylindrical fuel element.

Small displacements can be continuously and accurately measured in the reactor. For tensile tests on uranium during irradiation, Zaimovsky et al. (1958) used a linear differential transformer to measure the strain. Notley and Harvey (1963) applied an air gauge to the measurement of length changes of a Zircaloy-clad UO_2 fuel element during irradiation: They selected the method for its long-term stability and its insensitivity to irradiation damage.

Zaimovsky et al. stressed their specimens by pressurizing a pneumatic capsule from a gas supply whose pressure could be measured outside the reactor. The same principle has been modified by Reynolds (1963) and others to measure the pressure of fission gas within sealed fuel elements. A thin diaphragm at the end of the element separated the fission-product gas from an externally pressurized supply. When the diaphragm is in its undeflected position, as indicated by the making or breaking of an electric contact, the two pressures are equal and can again be measured outside the reactor.

Although wider application of these techniques could

greatly reduce the number of irradiations required, some of the other experimental difficulties should not be overlooked. In particular, appreciable variations during the irradiation in the neutron flux to which the specimen is exposed often make the results hard to interpret or even ambiguous. Many experimental and material-testing reactors are very poor with regard to the constancy of neutron flux at a given position. One must know, therefore, not only the mean values for flux-dependent variables, e.g., the fuel temperature, but also how they varied during the experiment.

Uranium- and Plutonium-Rich Alloys

2.1 Structures and Relevant Properties of Uranium Allotropes

Some of the properties normally associated with metals are a result of their simple, symmetrical structures. Any atom in the face-centred-cubic or hexagonal-close-packed structures is symmetrically surrounded by 12 equivalent atoms; in the body-centred-cubic structure by 8 nearest neighbours and 6 next-nearest neighbours. The outermost electrons in the electronic shells of each atom are not strongly localized between any particular pair of atoms and are therefore free to conduct electricity and heat, i.e., the conductivities are high. The high degree of symmetry, most notably in the face-centred-cubic structure, makes for several possible slip systems: This, and the absence of directed bonds, results in good ductility. Silver, iron, and magnesium are typical metals exhibiting these properties.

Elements classed as nonmetals, e.g., phosphorus and sulphur, form strong directed bonds with the outermost electrons localized between a few neighbouring atoms that are closer together than the others. Their open structures give a relatively low density, while the lack of free electrons results in low conductivities. The lower symmetry and the directed bonds cause poor ductility. Several elements, such as germanium and selenium, have intermediate properties and cannot easily be classified as either metals or nonmetals.

Uranium can exist in three different crystal structures, depending on the temperature range. Up to 660°C the stable form, alpha uranium, is orthorhombic. Such low symmetry, symptomatic of directed bonds, would not normally be expected to yield metallic properties. However, a detailed examination reveals that the structure is a slightly distorted version of the hexagonal-close-packed arrangement. Consequently, alpha uranium can be regarded as a metal, but one with lower than average conductivities and ductility. The high density (19.1 g/cm³) is partly derived from the near approximation to a close-packed structure. From 660 to 770°C the beta phase has a more symmetrical structure, tetragonal, but the large number of atoms per unit cell (30) reflects considerable complexity. The gamma phase, stable from 770°C to the melting point at 1130°C, is body-centred cubic. Accordingly, gamma uranium is notable for its high ductility.

Although some of the properties of alpha uranium derive from the similarity of its structure to a close-packed one, others can only be understood in terms of the departures from that structure. Most important is the anisotropy of many of the properties, i.e., the value of the property depends on the crystallographic direction. Linear thermal expansion provides a particularly striking example. From 0 to 200°C the coefficients are approximately 25×10^{-6}/deg C along the [100] direction, -0.6×10^{-6}/degC along [010] and 21×10^{-6}/deg C along [001]. The beta phase is also markedly anisotropic with coefficients of approximately 23×10^{-6}/deg C along [100] and [010], and only 5×10^{-6}/deg C along [001]. The cubic gamma phase is, of course, isotropic and has a coefficient of approximately 23×10^{-6}/deg C.

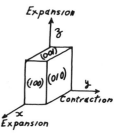

The properties of a polycrystalline specimen depend on the properties of the component grains and on their geometrical arrangement. Where a large number of grains are randomly oriented, the coefficient of linear thermal expansion is the same in all directions and has a value equal to the mean for the three principal directions in a single crystal. However, the specimen does not behave in all ways like an isotropic material: Along any arbitrary direction, the differential expansion of adjacent misaligned grains produces internal stresses, and even localized deformation, not present in isotropic metals. Almost any form of metal working causes a preferred orientation, or texture, in the arrangement of the grains. For instance, a rod of alpha uranium repeatedly cold-swaged with intermediate anneals can have an axial coefficient of thermal expansion around zero, which indicates that the {010} planes have become preferentially oriented normal to the axis.

Deformation of alpha uranium can occur by several mechanisms with the relative importance of the different modes depending on the temperature. Around room temperature, slip on {010} in the ⟨100⟩ direction is most common, with twinning on {130} composition planes next. The available choice of modes means that alpha uranium is reasonably ductile—elongations of 15% at room temperature are possible. Nevertheless, the distortions from a truly close-packed structure decrease the ability and ease of deformation. Thus, the hardness at room temperature is typically 250 VHN (cf about 50 for soft copper), decreasing with increase of temperature. On transforming into the beta phase the hardness rises, but only slightly, then drops to a very low value in the gamma phase. Uranium at 800°C is comparable to lead at room temperature.

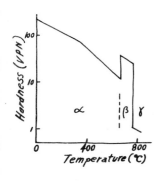

A property of considerable significance in determining the irradiation performance of a fuel is the thermal conductivity. That of randomly oriented polycrystalline alpha uranium is approximately 0.25 W/(cm deg C) at room temperature rising monotonically, but not linearly, to approximately 0.4 W/(cm deg C) at 600°C. These relatively low values (and corresponding ones for the electrical conductivity) reflect the tendency for electrons to be localized in the partially directed bonds of the distorted structure.

2.2 Dimensional Stability

Technologically, dimensional changes in the fuel are among the radiation effects that cause most concern. It is natural, therefore, that the various sources of dimensional instability in uranium metal have been studied extensively. Three separate phenomena, occurring during irradiation, can be distinguished:

1. Change in shape without the application of external stress and not associated with any appreciable volume change. This has been designated "irradiation growth".

2. Change in shape under an external stress. Such deformation is not confined to in-reactor service, but under irradiation the rates can be enhanced so much that "irradiation creep" merits separate consideration.

3. Increase in volume, initially without any major change in shape. This is called "swelling".

In addition, there are two other phenomena by which thermal cycling can cause dimensional instability in uranium:

1. If the fabrication of a polycrystalline specimen has produced a complex texture with at least two components, thermal cycling in the alpha phase can produce continuous elongation. Some of the experimental results and possible mechanisms have been summarized by Gittus (1963).

2. A polycrystalline specimen thermally cycled through the alpha-beta phase transformation suffers surface roughening, changes shape, and increases in volume by the development of internal cracks and pores. The effects are appreciable after as few as ten cycles. Buckley et al. (1959) have studied this form of instability.

Since these phenomena can be observed in the laboratory, they are not direct effects of irradiation. However, in most fuel irradiations the specimens are submitted to several, even many, thermal cycles. It is, therefore, most important to allow for the growth and swelling due simply to thermal cycling when interpreting the irradiation results. Further, the thermal cycles should be minimized if the irradiation effects are to be isolated.

2.3 Irradiation Growth

The change in shape without external stress, irradiation growth, was the first form of dimensional instability to be recognized. By the time of the first UN Conference on the Peaceful Uses of Atomic Energy at Geneva, at least three groups had already studied the phenomenon (Konobeevsky et al.; Paine and Kittel; Pugh, 1955). The reason for tremendous variations in growth behaviour between polycrystalline rods fabricated by different methods was understood, thanks probably to early experiments with single crystals. However, subsequent studies have shown that some of the comparisons made at that stage were misleading, owing to interference by factors whose importance was then unsuspected.

Figure 2–1 is the classic illustration of the effect of irradiation on a single crystal of alpha uranium. There was

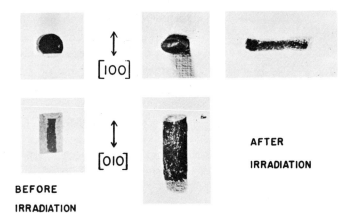

Fig. 2–1. Aspects of a cylindrical single crystal of alpha uranium before and after irradiation to 0.1 at.% burnup. (After Paine & Kittel (1955), courtesy United Nations.)

considerable elongation of the crystal in the [010] direction, contraction in the [100] direction by the same fractional amount, and no appreciable change in the [001] direction. Thus, the growth of polycrystalline rods was qualitatively explained: Those with a texture in which [010] predominated along the axis elongated during irradiation, while those with [100] contracted. In either case, compensating changes in the radial directions maintained the volume essentially constant. Also, since there is generally no preferred radial direction, the cross section remained round. The amount of growth for given irradiation conditions increased with the degree of axial texture in the rod.

The growth rate at any particular time appears to depend on the length of the specimen at that time, rather than on the initial length. Thus, the observations are generally analysed in terms of

$$L = L_0 \exp(Gf), \qquad (2\text{--}1)$$

where L_0 is the initial length, L the length after a fraction f of the atoms have fissioned, and G is the growth constant. Equation (2–1) is equivalent to

$$G = \frac{\log_e(L/L_0)}{f}, \qquad (2\text{--}2)$$

or, for small elongations, the approximation

$$G = \frac{\text{Percentage   Growth}}{\text{Percentage   Burnup}}. \qquad (2\text{--}3)$$

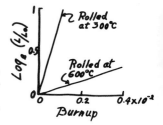

Although Kittel and Paine (1958) have confirmed Eq. (2–1) for certain texture polycrystalline specimens, the growth of single crystals has not been studied systematically as a function of burnup.

The value of G is used to compare the growth of different specimens, of similar specimens under different conditions, or of a single crystal in different directions. When G in the [010] direction (G_{010}) equals 500, which is in the range commonly experienced, each fission event has had the net effect of transferring 500 atoms from {100} to {010} planes, regardless of the mechanism. On the macroscopic scale, a burnup of only 0.2% (1850 MWd/tonne) would result in an elongation of approximately 100%.

After years of controversy, the effect of the irradiation temperature on the growth constant has now been clearly demonstrated in independent series of experiments. Buckley (1964) has compiled results (reproduced as Fig. 2–2) that show the constant to be very large at low temperatures, relatively independent of temperature in the range 0 to 300°C, and much smaller at higher temperatures. Further work by Buckley (1965) has confirmed the validity of the experimental points for polycrystalline specimens deformed 58% at 300°C (denoted by ○ in Fig. 2–2). Thus, there is a real departure from the smooth curve shown and the decrease in growth rate at high temperatures occurs in two stages.

The relationship of Fig. 2–2 includes, or is consistent with, all but one of the available experimental results. Loomis et al. (1964) reported that a slightly imperfect uranium single crystal exhibited relatively little growth ($G_{010} = 150$) when irradiated at $- 163$°C. Although the low-temperature exposure was terminated after only 41 h, the final burnup, 1.1×10^{-6} of the uranium, and the rate

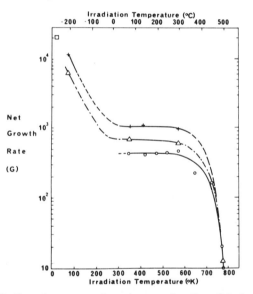

Fig. 2–2. Growth constant of uranium as a function of the irradiation temperature. (After Buckley (1964).).

○ Polycrystalline, 58% deformation at 300°C, burnup 0.03 to 0.06% in 100 days. Each point is mean of 23 specimens.

△ Polycrystalline, 58% deformation at 20°C, burnup 4×10^{-6} in 14 days. Each point is mean of 4 specimens.

+ Single crystal, measured parallel to [010], burnup 2×10^{-6}. Each point is mean of 2 specimens.

□ Polycrystalline, strongly textured. (After Quéré & Doulat (1961).)

of burnup were both comparable to the equivalent values for specimens that showed much larger growth. Loomis (1965) has since exposed the same specimen to further irradiation: After periods of irradiation at 165 to 200°C, G_{010} subsequently measured at -140 to -170°C was more than tenfold greater than previously, and the values at all temperatures were in agreement with those compiled by Buckley.

The first result by Loomis et al. appears to show that there can be an incubation period for growth under certain conditions, provided there was no malfunctioning of their apparatus which measured length changes during irradiation. A demonstration of the reproducibility of the observation seems desirable.

Several variables apart from irradiation temperature may affect the growth. For temperatures around 470°C, Buckley is reported to have shown that increasing the rate of burnup increases the growth constant, and hence extends the range of appreciable growth to higher temperatures (Pugh, 1964). The effects on growth of grain size and impurities have not been adequately established: Some comparisons have suggested that growth is greater in fine-grain material, but it was not demonstrated that the texture and impurity content remained constant. Buckley (1961) showed that plastic deformation of single crystals or highly textured polycrystalline specimens, by as much as 25%, had no effect on the growth either at 75 or -196°C. Weinberg and Quéré (1962), however, found higher values of G_{010} in heavily cold-worked than in annealed material: The factorial increase was about 2 at -252°C, 3 at -196°C, and 17 at 40°C. Plastic deformation that resulted in a change in the texture would, of course, cause a change in the net growth of a polycrystalline specimen.

The growth of a polycrystalline rod is related to its crystallographic texture, but it would be wrong to take the vector sum of the growth of the component grains, since the individual deformations generate mutual restraints. Buckley (1962) demonstrated experimentally that an external stress did not significantly affect the growth of single crystals. Strip specimens were stressed in bending under four-point loading while being irradiated at −196°C. If tensile and compressive stresses affected the growth in opposite senses, the specimen's curvature should have changed markedly. However, the changes observed were very little different from those in laboratory control specimens, despite an elongation of about seven percent due to growth in the irradiated ones.

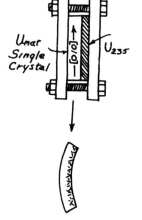

Complete restraint can result in plastic deformation within the crystal. Buckley clamped together in a restraining frame a thin single crystal of natural uranium and a foil of highly enriched uranium. During irradiation, fission fragments from the foil caused most of the damage to occur in a layer on one side of the single crystal. Since the [010] direction lay along the length of the specimen the layer would have elongated in its own plane, had it not been subjected to restraint. After removal of the enriched foil the specimen was again irradiated, now free of restraint. This time the density of damage was uniform but the specimen became curved, due to the previously damaged layer elongating less rapidly than the remainder, by as much as 20%. This was presumably because the restrained growth broke up the single crystal in the damaged layer, which rotated regions of the lattice from their original orientation: Slip and twin traces were apparent on the surface of the damaged layer. Thus, restraint does not prevent the growth of alpha uranium but causes associated plastic deformation to maintain the dimensions constant.

Buckley (1962) went on to show how the growth coefficient G depends on the degree of texture in the sample. Wires were prepared with varying amounts of {010} texture by swaging to different reductions. Measurements of the coefficient of axial thermal expansion were interpreted in terms of the fraction of {010} planes with normals parallel to the axis: It was known that the coefficients for random aggregates and for a pure {010} texture are 14.6×10^{-6} and -0.7×10^{-6} /deg C, respectively, over the appropriate temperature range. The results (see

Fig. 2–3) indicate that polycrystalline aggregates with an alignment of 80% or over behave essentially as single crystals. For comparison, a straight line between the extremes of the curve would represent the approximation that the net growth equals the vector sum of the growth of component grains.

To maintain dimensional stability of a fuel element, the net growth of a uranium rod can be reduced to negligible proportions by providing a random texture, e.g., by quenching the uranium from the beta phase. The consequent plastic deformation in adjacent grains growing in opposing directions is clearly seen at low burnup from metallographic sections (Sec. 2.9). At higher burnup, however, the structure is so heavily deformed that the detail is practically unresolvable. Much experimental work has been done on correlating the growth coefficients with variations in the fabrication process, and a great deal of empirical information is available. Although the observations can be generally explained by the texture of the samples, it has not yet proved possible to make exact predictions of the growth of polycrystalline rods. Buckley's results of Fig. 2–3 provide the necessary information, but only for the specific conditions of his experiment. Since restrained growth produces lattice rotation, the degree of texture will vary as growth proceeds. In a rod with

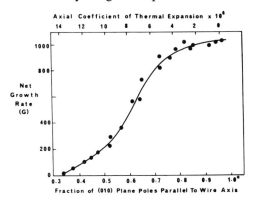

Fig. 2–3. Variation in the net growth rate of polycrystalline uranium wires with the fraction of (010) plane poles aligned parallel to the axis. (After Buckley (1962), courtesy Inst. of Metals.)

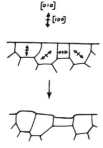

a strong {010} texture, growth of the favourably oriented grains will tend to force the unfavourably oriented grains into alignment and, hence, increase the growth rate. Furthermore, a texture is not uniquely determined by the fraction of {010} planes with normals parallel to the axis; at least one other component must be specified.

Even when a random texture prevents net elongation, severe localized distortions can occur at the surfaces of the rod. Surface grains, being free from restraint on one side, deform according to their orientations. Thus, those whose [010] direction is normal to the surface will form raised bumps, while those whose [100] direction is normal will form depressions. The overall effect is of surface roughening, as illustrated in Fig. 2–4. Figure 2–4b shows how a

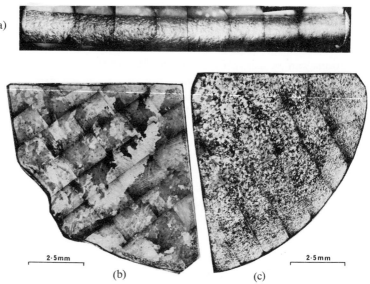

(a)

(b)   (c)

Fig. 2–4. Surface roughening in irradiated uranium.
(a) Composite photograph of rod with sheath stripped off showing variation in roughening along the length.
(b) Photomicrograph of a quadrant showing a surface bump due to a large grain in the interior −600 MWd/tonne at 200°C.
(c) Photomicrograph of a quadrant showing greatly reduced surface roughening with fine-grain uranium −500 MWd/tonne at 230°C.
(After Eldred et al. (1958), courtesy United Nations.)

particularly large grain, although relatively deep below the surface, can cause a large bump. The phenomenon of surface roughening by growth is variously referred to as "orange peeling" and "wrinkling". Konobeevsky et al. (1955) demonstrated that a sufficiently strong sheath could restrain the uranium's growth and, hence, prevent roughening. An annulus of uranium was filled with a soft magnesium cylinder and irradiated in a steel sheath. Afterwards, the inner surface of the uranium was found to have roughened, but not the outer one. In practice, a strong sheath is not usually an acceptable remedy, owing to the consequent parasitic neutron absorption. Rather, the amplitude and wavelength of the roughness are decreased by employing uranium of small grain size. Figure 2–4c illustrates the benefit of fine-grain uranium. Zaimovsky et al. (1958) showed a quantitative correlation between the surface roughening and burnup, with the greater roughening occurring in specimens whose heat treatment had given them a large grain size.

Eldred et al. (1958), from observations on many uranium fuel elements, concluded that roughening was most serious for fuel temperatures in the range 150 to 200°C. Since roughening involves plastic deformation as well as growth, this conclusion is still consistent with the growth increasing rapidly with decreasing temperature, below 0°C. Difficulty in deforming adjacent grains would explain the absence of roughening at low temperatures, while decreasing growth would explain its absence at high temperatures.

Other anisotropic, but nonfissionable, metals exhibited pronounced growth when bombarded by fission fragments (Buckley, 1961). At first, it seemed that the concentrated damage of fission spikes was necessary for growth: Thompson (1961) bombarded a uranium specimen with 3-MeV protons to a total energy deposition that would have given appreciable growth, had it been produced by fission fragments, but no dimensional changes were observed. Later, however, Buckley appears to have obtained growth on bombarding alpha uranium with 50-keV protons

(Pugh and Butcher, 1964). Any changes in the uranium lattice cell dimensions are negligible compared with changes in the crystal dimensions due to growth.

The many theories of growth that have been advanced can be broadly divided into two groups. In one, thermal expansion of the fission spike stresses the surrounding matrix, causing plastic deformation. The anisotropic mechanical properties of uranium result in a net elongation in the [010] direction, with different modes of deformation operating during expansion and contraction to explain the lack of reversibility. In the other group are theories that depend on the interstitial atoms and vacancies from the fission spike returning to the lattice along different preferred directions, or to preferred planes. The large and increasing growth at low temperatures where diffusion is reduced is a strong argument against directional diffusion of vacancies and interstitials being the important mechanism. The insensitivity of growth to prior plastic deformation suggests that dislocations are not important as sinks for point defects, and it also renders the plastic deformation models less likely.

The most reasonable explanation is due to Buckley (1961). Theories of fission spikes (Sec. 1.2) have suggested that the vacancies will be concentrated in the core of the spike, while the interstitials will be found mainly in a surrounding shell. This situation presents an opportunity for the two types of point defects to condense into separate dislocation loops, rather than mutually to annihilate each other. The anisotropic stresses, due to thermal expansion of the spike, provide a reason for the vacancy and interstitial loops condensing on different crystallographic planes. Buckley argued that in alpha uranium the interstitials should condense on {010} planes and the vacancies on {110} planes, thereby explaining the observed growth.

Transmission electron microscopy confirmed the basis of the explanation but showed that the detailed mechanism should be somewhat modified. Makĩn et al. (1962) clearly observed dislocation loops in irradiated uranium (Sec. 2.9). However, their analysis of the data has since been

corrected by Hudson (1964). The latter identified dislocation loops on {010} and {100} planes, with Burgers vectors
$\frac{1}{2}\sqrt{a^2 + b^2}\langle 110\rangle$ and $a[100]$, respectively. If it is assumed
that the former are composed of interstitials and the latter
of vacancies, the observed growth is explained. Following
irradiation to a fractional burnup about 10^{-6} of the
uranium at $-196°C$, the loops were randomly dispersed;
at 80°C they had begun to align into rows; and at 350°C
the alignment was complete.

Loops that formed during irradiation at low temperature
annealed rapidly at temperatures above 500°C which
explains the decrease in growth in this range. No loops
were observed in Thompson's proton-bombarded uranium
that exhibited no growth, while loops were observed in
all Buckley's specimens that exhibited growth — the
nonfissionable specimens, his proton-bombarded uranium,
and the neutron-irradiated uranium. Thus, dislocation
loops in anisotropic materials appear to be associated
with growth and are presumed to be a cause of it. Further
discussion of the other mechanisms that have been proposed
to explain growth is to be found in Buckley's paper (1961).

2.4 Irradiation Creep

Irradiation creep provides an excellent example of theory
leading to the discovery of an effect and of the misunder-
standings introduced by translation between languages.
There is little doubt that an acceleration due to irradiation of the creep of alpha uranium was first observed
in Russia, prior to the first UN Conference at Geneva.
Konobeevsky et al. (1955) reported the fact, but the
magnitude of the effect was not apparent from reading
the English translation of their paper. When clarification
was sought during discussion at the conference, G. S.
Zhdanov, on behalf of the authors, gave a reply which
appeared in the English version of the proceedings as:

> (There is) "an increase in uranium creep during
> irradiation of the order of 1.5–2, and not 1.5–2
> percent."

This was generally taken to mean an increase by a *factor* of approximately 1.5–2. It was not until the Second Geneva Conference that Zaimovsky et al. (1958) stated that the interpretation should have been 1.5–2 *orders of magnitude*, i.e., a factor of 50 to 100.

In the intervening period, Cottrell had been studying the factors affecting swelling of uranium (Sec. 2.5). He reasoned that in polycrystalline alpha uranium under irradiation, where the growth of adjacent grains is causing plastic deformation, the internal stresses must be high and continuously regenerated. Under these circumstances a small external stress could have a relatively large effect, i.e., there could be appreciable creep rates at temperatures low enough for thermally activated creep to yield negligible rates. To test the possibility he and his colleagues devised an elegant experiment. Helical springs wound from natural uranium wire were subjected to a dead load and sealed in transparent capsules (Roberts and Cottrell, 1956). After each week's irradiation the capsules were withdrawn from the reactor to permit remote measurement of the extension of the spring optically. In a thermal neutron flux of 10^{12} n/(cm^2 sec) and at a temperature of $100°C$, appreciable creep was produced by stresses only about one percent of the normal yield strength.

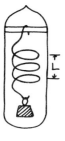

Roberts and Cottrell presented an order-of-magnitude analysis of the phenomenon. Starting with annealed polycrystalline material, the intergranular stresses would build up to the yield strength σ_y in a time t_m given by

$$t_m \approx \frac{\sigma_y}{E\dot\varepsilon_g},\tag{2–4}$$

where E is the elastic modulus and $\dot\varepsilon_g$ the growth rate of an individual crystal. Thus, there should be an incubation period t_m before irradiation creep becomes apparent, and the same interval should be required for relaxation of any externally applied stress. An external stress σ would produce an elastic deformation σ/E. If stress relaxation contributes a strain increment σ/E in each time interval t_m the creep rate is

$$\dot{\varepsilon}_e \approx \frac{\sigma}{Et_m} \approx (\sigma/\sigma_y)\dot{\varepsilon}_g. \qquad (2\text{--}5)$$

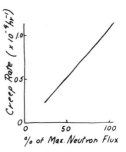

Naturally, the deformation does not occur in this step-wise manner, but Anderson and Bishop (1962), using a more sophisticated analysis, deduced the same functional relationship. The irradiation creep rate should be proportional to the external stress and to the growth rate of a single crystal, hence to the neutron flux for a given enrichment of uranium.

In practice, Roberts and Cottrell found an incubation period of 6 ± 1 days, compared with a prediction of about one week. After two weeks of transient creep the magnitude of the steady-state rates, and their dependence on external stress, were also in good agreement with the predictions. With specimens under uniaxial tension, Zaimovsky et al. (1958) confirmed the dependence of the irradiation creep rate on external stress and demonstrated that the rate was proportional to the flux. However, these authors commented on the absence of an incubation period in their experiments, although application of the Roberts and Cottrell relationship to their conditions predicted an observable effect. At $-196°C$, Buckley (1962) measured very rapid irradiation creep in poly-crystalline specimens bent under four-point loading; about ten times the rate observed by Roberts and Cottrell for comparable stresses at $100°C$. This increase supports the analysis of Eq. (2–5) since Buckley (1961) had earlier shown a similar increase in the growth rates at the lower temperature. (Roberts and Cottrell would not have predicted this result because, at that time, they believed the growth rate to be negligible at $-196°C$.) Although Buckley's presentation shows no incubation period, the predicted period of about half a day is consistent with his results.

Konobeevsky et al. (1958, 1960) studied stress relaxation during irradiation at about $150°C$ in U-0.91wt%Mo (alpha phase) and U-9wt%Mo (gamma phase). A flat specimen was first bent into an arc and then irradiated while clamped between flat plates: The change in curvature

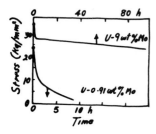

as a result of irradiation was interpreted in terms of stress relaxation. In both alloys there was a rapid change in the original stresses during the first five hours of irradiation but, whereas the stress relaxation was almost complete in the alpha phase, only a small fraction of the stresses disappeared in the gamma-phase alloy even after 96-h irradiation. Post-irradiation annealing of either specimen showed a reversal of the relaxation, with the original internal stresses being partly restored. Observations by x-ray diffraction gave closely parallel results: Line broadening in worked samples of the same alloys was reduced by irradiation, by a greater amount in the alpha phase than the gamma phase and in about the same time. Again, subsequent annealing produced a partial reversal of the effect.

Most of the stress relaxation observed by Konobeevsky et al. occurred within a period short compared with that necessary for the internal stresses in individual grains to build up to the yield point by the irradiation growth mechanism. Thus, although these stress relaxation effects may have contributed to some of the transients in the experiment of Roberts and Cottrell, they are most unlikely to have played any part in the continuing accelerated creep. Konobeevsky and his colleagues proposed a mechanism for the stress relaxation that also explained the reversible nature of the phenomenon. They supposed that, as a result of the stresses, fission spikes would leave behind them more vacancies than interstitials in regions of compression (and conversely for regions in tension). The stresses would be relaxed but, since the concentration of vacancies would be greater than that in thermal equilibrium, subsequent annealing without irradiation would remove the excess vacancies and restore the stresses. They also proposed a mechanism for stress relaxation that would not be reversible: Irradiation-accelerated diffusion might cause a to those in of atoms from the regions in compression net transfer tension.

Yet another possible mechanism to explain irradiation-accelerated creep was proposed by Hesketh (1962). He accepted the explanation of irradiation growth due to Buckley and Makin et al. (Sec. 2.3), but pointed out that the observed direction of growth was due to a balance between the number of interstitials (and similarly for vacancies) forming loops on two equivalent planes with two equivalent Burgers vectors. The presence of stress could, however, destroy the equivalence and rotate the direction of positive growth away from [010]. Buckley's finding that stress had no effect on growth (see Sec. 2.3) does not contradict this argument. His experimental conditions, with the [010] axis of the specimen along the neutral axis for the bending stresses, constituted a special symmetrical case where no effect would be expected. However, Hudson has since shown that neither type of loop forms on two equivalent planes (Sec. 2.3).

The mechanisms proposed for irradiation-accelerated creep of uranium can be divided into two categories: those that depend on the anisotropic properties of alpha uranium, and those that do not. Konobeevsky et al. showed that the difference in structure had a large effect on stress relaxation, but creep may be associated with a separate mechanism. It would, therefore, be most useful to have a determination of the creep rates under irradiation of a gamma-phase (cubic) alloy for comparison with those for alpha uranium already studied. In completely different materials, greater plasticity during irradiation was observed in anisotropic $ZrO_2 - UO_2$ than in isotropic UO_2 (Sec. 3.12).

Rose (1957) measured the creep rate during irradiation of polycrystalline alpha uranium under compression at 450°C, i.e., at a temperature where growth is negligible. Although no large change in the secondary creep rate was observed, the possibility of an irradiation-induced component was not eliminated: The thermal creep at that temperature is relatively rapid and the experimental scatter was appreciable. The experiment had been designed to test whether there was an increase by orders of magnitude, as required then to explain certain observations on swelling. A fresh stage of primary creep occurred after each reactor shutdown. Rose attributed this to thermal-cycling stresses, but the stress-relaxation phenomenon noted by Kono-beevsky et al. could also be partly responsible.

Acceleration of the creep of uranium during irradiation can have a very real significance to reactors. For instance, where fuel elements are stacked vertically, one on top of another, the dead load imposed on the lower elements can produce serious bowing stresses. If the creep strength is insufficient to prevent the bowing, a simple remedy is to reduce the unsupported length. Cottrell's theoretical analysis predicted bowing of the elements in the Calder Hall reactor; this permitted the preparation of mid-point braces, to be fitted to the elements when bowing was actually observed.

2.5 Swelling

When one atom of uranium fissions, two new atoms are produced, and on an average each is larger than the original atom. Howe and Weber (1957) calculated that the percentage increase in volume should be approximately three times the percentage burnup. The exact value depends on the valencies deduced for the various fission products, and not all these are well established for their ill-defined environment. Ignoring the small number of atoms in interstitial sites and any initial porosity that may be taken up, this estimate provides the minimum volume increase to be expected of irradiated fuel. Any changes in lattice parameter are small and can usually be neglected. Where experimental values of swelling are quoted, they usually refer to the total volume increase.

In practice, large increases can be observed in irradiated fuel; sometimes over 100% of the initial volume. The difference from the minimum estimate is accounted for by voids and porosity distributed through the solid (Fig. 2–5). Since about 15% of the fission-product atoms are the rare gases xenon and krypton*, the pores are probably

---

* The amount generated by 1% burnup of 1 cm³ of uranium would occupy roughly 5 cm³ at standard temperature and pressure.

gas-filled. In uranium specimens bombarded with natural xenon and subsequently annealed, electron microprobe analysis has shown the resulting pores to be xenon-filled (Levy et al., 1961). The large swellings, associated with porosity, were first reported by Pugh (1955). Earlier, the volume increases had corresponded closely to the minimum expected (Paine and Kittel, 1955). Subsequently, the existence of swelling has been adequately confirmed, but the dependence on potential variables such as temperature, burnup and structure are even now not fully established. One reason for this unfortunate situation is that many factors are probably important — one author listed 21. In fuel irradiations, it is very difficult to control some of these variables, especially the fuel temperature, to the required accuracy. Also, the tremendous effect of impurities on the swelling of alpha uranium was not appreciated for a long time (Sec. 2.7).

Swelling is probably the most universal problem encountered in fuel irradiations. Although first recognized in uranium metal, swelling has since been observed in all types of fuel, including ceramics such as UO_2. It also occurs in nonfissile materials where neutron capture yields a rare-gas product, e.g., the helium produced by irradiation in boron-containing control materials or in beryllium-containing moderators. Since it is in connection with uranium metal that theories to account for swelling have advanced furthest, considerable space is devoted to the subject in this section.

To permit comparison of swelling between specimens irradiated to different burnup, the volume increase is often normalized to a burnup of one percent, implicitly assuming a linear relationship. For volume increases up to ten percent, there is some experimental support for this assumption (Pugh, 1961; Bentle, 1961), but the scatter in the experimental results is too great to establish the dependence conclusively; Bellamy (1962) showed for his specimens that a linear relationship was only an approximation and suggested that the volume increase associated with the porosity was proportional to the burnup raised

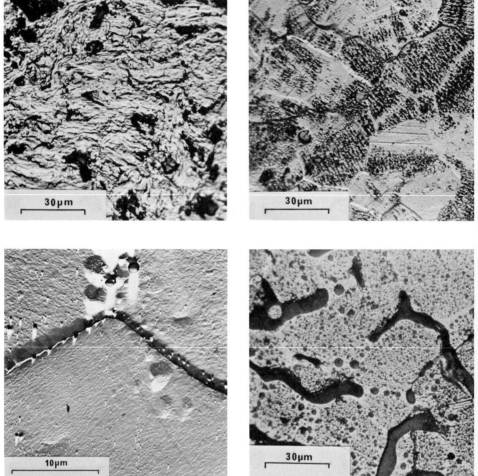

(a)   (b)

(c)   (d)

Fig. 2–5. Swelling in irradiated uranium. Voids appear black; in (c) the "shadows" of voids appear white.

(a) Tearing and deformed microstructure — 1400 MWd/tonne at 425°C.

(b) Pores aligned along crystallographic directions in the grains — 280 MWd/tonne at 520°C.

(c) Electron micrograph showing grain-boundary porosity and extremely fine pores throughout the grains — 930 MWd/tonne at 625°C.

(d) "Breakaway" swelling with pore segregation at the prior grain boundaries.

(After Leggett et al. (1964), courtesy United Nations.)

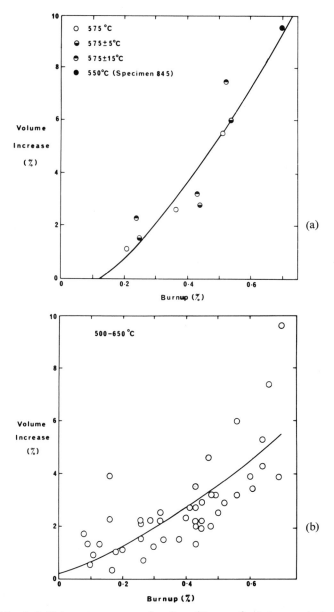

Fig. 2–6. Volume increase as a function of burnup for beta-quenched
uranium with iron and aluminum additions.

(a) Temperatures as shown. (After Bellamy (1962), courtesy Inst. of
Metals.)

(b) Temperatures 500 to 650°C. (After Bellamy (1965).)

to the power 3/2 (Fig. 2–6a). Recent results by the same author (Bellamy, 1965) indicate a more nearly linear relationship for specimens relatively free of the grain-boundary cracks that had been prevalent in his earlier specimens (Fig. 2–6b). For volume increases over roughly ten percent, there is a much more rapid variation with burnup, so that this stage is sometimes referred to as "breakaway swelling". Examination of the specimens that have suffered breakaway swelling usually reveals large, irregular, and irregularly distributed voids (Fig. 2–5).

Just recently, there has been a major clarification concerning the effect of the irradiation temperature on swelling. It now seems that much of the earlier confusion resulted from a failure to appreciate that the swelling versus temperature relation can be profoundly affected by other factors. Figure 2–7 compares the results published simultaneously by two independent groups (Kittel et al.; Barnes et al., 1964). (The relevant observations summarized by the first group appear in a report by Leggett et al., dated November 1963 but not publicly released until November 1964). Most significantly, both groups found that a very high maximum can, under certain circumstances, occur in the swelling versus temperature curves around 450°C. Kittel et al. observed the maximum in unrestrained specimens of normal purity, but not in large, sheathed specimens. Furthermore, no maximum occurred for specimens subjected to post-irradiation annealing after exposure either cold or restrained. Barnes et al. encountered the maximum in small, unrestrained specimens operating at high fission rates (above 15 MW/tonne) but not at low (about 4 MW/tonne): The swelling maximum occurred at lower exposures for material of lesser impurity content. (The effect of impurities is considered further in Sec. 2.7.)

Both groups associated the large volume changes with ragged tears in the structure. While Barnes et al. noted only intergranular tearing, Adda et al. (1964), who also observed the enhanced swelling in the same temperature range, found that twin boundaries, too, acted as sites for the cracks. The most thorough metallographic study of

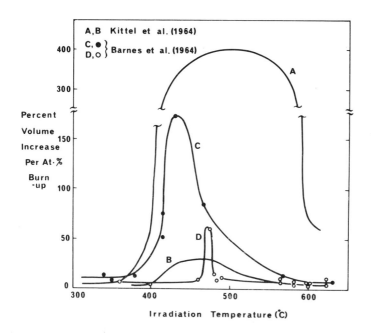

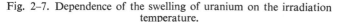

Fig. 2–7. Dependence of the swelling of uranium on the irradiation temperature.

A — High purity uranium.
B — Normal purity uranium.
C — "Adjusted" uranium, 6000 to 8000 MWd/tonne.
D — "Adjusted" uranium, 4000 to 5000 MWd/tonne.

the phenomena was performed by Leggett et al. (1964). For irradiation temperatures below 350°C the structure shows evidence of severe working but there is negligible swelling: From 400 to 500°C tears along grain boundaries give very large swelling. The stresses causing the tears probably result from opposed irradiation growth in misaligned grains, with the dependence on fission rate due to the growth being dependent on that variable (Sec. 2.3). Barnes et al. supposed that where growth occurred at sufficiently high temperatures, deformation would take

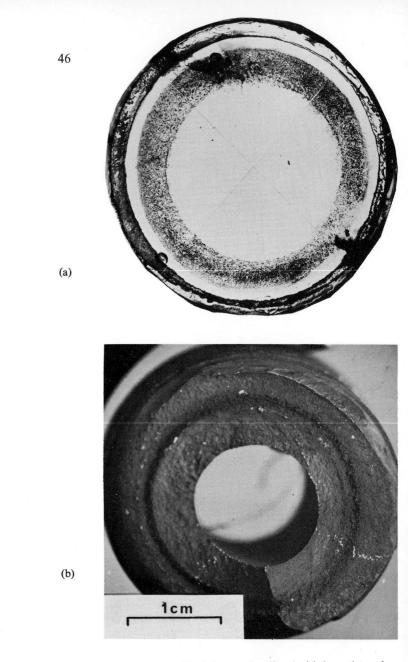

(a)

(b)

Fig. 2–8. Cross sections of fuel elements irradiated with large thermal gradients and low external pressure, showing a porous annulus at a specific temperature.

(a) Annulus at about 400°C in "adjusted" uranium. (After Stewart et al. (1964), courtesy United Nations.)

(b) Annulus at about 420 to 450°C in U–1wt%Mo. (After Englander et al. (1964), courtesy United Nations.)

place by grain-boundary sliding and cavitation in preference to plastic flow within the grains. Diffusion of fission-product gases into the voids could accelerate the grain-boundary swelling and tearing. In accord with this explanation, Adda et al. observed the tearing to be much less severe in highly textured specimens.

Thus far, there is general agreement on both the observations and the interpretation. From 500 to 600°C, however, Leggett et al. still measured large volume increases in their uranium, which was of specially high purity, while Barnes et al. did not (Fig. 2–7). In this range, the swelling is caused by large pores or possibly tears aligned along crystallographic directions within the grains (Fig. 2–5b). Just above 600°C, there is again agreement, with small volume increases attributed to pores at grain boundaries and many tiny pores (about 0.1-μm diam) throughout the grains (Fig. 2–5c). The broad peak of Fig. 2–7, therefore, consists of two unresolved peaks, reflecting the fact that at least two mechanisms exist that can produce exaggerated swelling between 400 and 600°C. The third type of swelling, due to homogeneously distributed gas-filled porosity and for which most of the theories have been developed, does not normally become significant in uranium until temperatures over 600°C are reached.

Graphic demonstration of the sharp maximum in the swelling is obtained from cross sections of large rods. Stewart et al. (1964) observed a porous annulus corresponding to temperatures around 450°C (Fig. 2–8), while Englander et al. (1964) noted a similar effect for the region between 420 and 450°C in U–0.5wt% Mo and U–1.1wt% Mo alloy rods. These observations indicate that the self-restraint of massive specimens is not alone sufficient to prevent the exaggerated swelling. Obviously, the volume changes due to the localized tearing can be serious and certainly must be allowed for in analysing the changes attributable to uniformly distributed porosity.

Some of the earlier results can now be more easily understood. Re-analysis by Granata and Saraceno (1963) of data published by Pugh (1961) yields a similar, if smaller, maximum in the swelling at about the same temperature. Greenwood (1959) observed variable volume increases up to 150% for specimens believed to have been irradiated at moderate temperatures within the alpha phase. From a detailed analysis of the thermal history of his specimens, Greenwood suggested that large swelling occurred where there had been many thermal cycles. Subsequent examination (Pugh, 1961) amplified this finding. The effect of the thermal cycling was apparently confined to the formation of grain-boundary cracks in the alpha phase, with no significant change in the distributed porosity. Pugh suggested that film boiling might have occurred in the static sodium surrounding Greenwood's highly rated specimens, causing substantially higher fuel temperatures than those calculated from readings of thermocouples in the sodium. Metallographic observations of spherical porosity, rather than ragged tears, in these specimens supported the suggestion of high temperatures. However, thermal cycling and periods of irradiation in the temperature range for growth may also have contributed to the large swelling. Kittel and Paine (1958) presented results that showed low swelling up to 430°C and then a rapid increase with temperature up to 550°C, the highest temperature tested. Had these investigators continued to higher temperatures, they could conceivably have passed through a maximum swelling.

The existence of several different mechanisms for swelling must be borne in mind when assessing the effect on the volume increase of particular variables and when testing theoretical models. Thus, a given variable may affect one mechanism much more than others. For instance, the swelling has been seen to increase more rapidly with burnup when grain-boundary tearing is prevalent.

Many of the measurements of swelling have been obtained from small specimens irradiated with only small temperature differences between the surface and the centre. With large uncertainties in the effect of the temperature on swelling when only one temperature is involved, it is not

yet possible to predict the swelling of practical fuel elements
having large temperature differences. The extent of the
ambiguity was presented forcibly by Bentle (1961). Figure
2–9 shows how he plotted the same set of experimental
results for beta-quenched uranium in two different ways;
as a function of either the surface or centre temperature.
Judging from the amount of scatter in each curve, the
correlation with the surface temperature may be the more

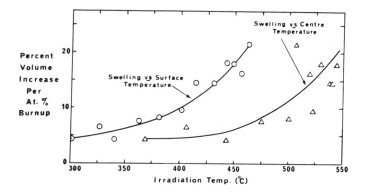

Fig. 2–9. Swelling of cast and beta-quenched uranium plotted
as a function of either the surface or centre temperature. (After
Bentle (1961).)

significant one, but specific investigation of this aspect
seems desirable. Regrettably, some investigators do not
state explicitly which temperature they intend when they
quote a "fuel temperature".

The swelling of uranium, up to a volume increase of
about ten percent, has been inferred to be insensitive to
the external pressure, but there has been no clear experimental demonstration under irradiation. If it is assumed that virtually all the rare gas generated by fission
is in spherical pores, application of the general gas equation

provides the pressure in the pores (p) as a function of the fractional volume increase ($\Delta V/V$). Thus,

$$p\Delta V/V = Bk\Theta/a^3, \qquad (2\text{-}6)$$

where k is Boltzmann's constant and B the fraction of gas atoms per atomic site of volume a^3 and at absolute temperature Θ. A more exact analysis uses Van der Waal's equation to describe the behaviour of the gas in the pores (Barnes, 1964). For a burnup of one percent causing a volume increase of ten percent at 500°C, the calculated pressure in the pores is about 150 atm (Fig. 2–10). Where the swelling is due to uniformly distributed porosity, therefore, an external pressure would not be expected to have any appreciable effect, if its magnitude is small compared with that calculated for the gas in the pores. However, where grain-boundary tearing is the predominant mechanism, external restraint may be much more effective in reducing swelling.

Makin et al. (1962) reported that an unrestrained uranium specimen irradiated at 530°C decreased in density by 8%, while a restrained one exposed under otherwise similar conditions decreased by only 0.5%: Other specimens that were free to expand a little before becoming restrained suffered intermediate density decreases. Metallography, however, revealed that these specimens contained large grain-boundary voids (over 1-μm diam.). At these sizes the effect of surface tension is relatively small and the external restraint can be important.

The effect of the external pressure on swelling, during the post-irradiation annealing of uranium that had been irradiated cold, was investigated by Churchman et al. (1958). They found that application of a sufficiently high pressure could decrease the amount of swelling and, moreover, that the volume changes were nearly reversible with changes in the pressure. In accord with the simple analysis of Eq. (2–6), an external pressure of 400 atm was required to influence the swelling of specimens in

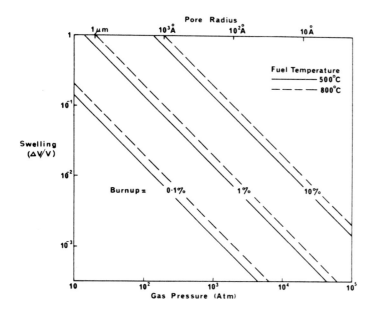

Fig. 2–10. The calculated swelling as a function of pore radius or gas pressure in the pores, for selected values of the burnup and fuel temperature.

which the gas pressure was calculated to be several hundred atmospheres. The reversibility of the changes suggests that the pores are gas-filled. Some confirmatory evidence is obtained from a correlation between the fractional release of fission-product gas from the uranium and the volume increase. Up to approximately 20% increase, when the individual pores are closed, the gas release is very small. It is only with larger volume increases, when the pores become interconnected and open to the outer surface, that the release is appreciable. Samples that have doubled in volume normally release over half their gas.

When a pore of radius r is in equilibrium

$$p = p_e + p_i + \frac{2\gamma}{r}, \qquad (2\text{-}7)$$

where p_e is the external pressure on the fuel, p_i the pressure due to internal stresses in the fuel, and γ the surface tension of the material surrounding the pore. Equation (2-7) shows that the external pressure should become increasingly important as the pores coarsen. Differential thermal expansions between the hot centre and the cooled surface of the fuel result in p_i having a positive value at the centre and a negative one near the surface. In alpha uranium, irradiation-enhanced creep and stress relaxation (Sec. 2.4) will rapidly reduce the magnitude of p_i to a value comparable with the external pressure. Even in isotropic fuels, very small differences in the amount of swelling between centre and surface would suffice to relieve the thermal stresses. Thus, until the pore becomes large, the surface-tension term is most important in balancing the gas pressure.

The conditions under which the external and internal pressures can be neglected are determined by comparing the values of p_e and p_i with the gas pressure obtained from Fig. 2-10. The radius of a pore in equilibrium with the pressure along the abscissa, if p_e and p_i are put equal to zero in Eq. (2-7), is shown along the top of the figure. A value of 10^3 dyn/cm has been assumed for γ, but, since the surfaces may be contaminated by fission products, there is little justification for this particular value. Where the pores are not all of the same radius, but are represented by a distribution of n_i pores of radius r_i per unit volume an average pressure can be obtained as

$$\bar{p} = \frac{2\gamma \, \Sigma \, (n_i r_i^2)}{\Sigma \, (n_i r_i^3)}. \qquad (2\text{-}8)$$

To test the relation, Greenwood (1962) selected for metallographic examination specimens that had exhibited substantial swelling. From counts of the pores and measurements of their radii, he computed the values of $\bar{p}$, the restraining pressure due to surface tension. The good agreement with the gas pressure calculated from the volume increase (Eq. 2–6), for all eight specimens compared, gave support to the initial assumption.

The tentative conclusion that the gas pressure in the uranium pores is balanced by surface tension is identical to that reached by similar methods for rare gas in non-fissile metals, e.g., helium in copper and aluminum (Barnes, 1959; Boltax, 1962; and Barnes and Mazey, 1963). While studying helium in copper, Barnes and Mazey (1963) observed the pores by transmission electron microscopy. The absence of dislocations associated with pores suggested that plastic deformation was not occurring around the pores and, hence, that their internal pressure did not greatly exceed the restraining pressure. Similar observations have been made on pores in irradiated uranium (Makin et al., 1962). To maintain the equilibrium when the amount of gas in a pore increases, an appropriate number of vacancies must diffuse into the pore. In all these studies the method of comparison is sufficiently crude that departures from true equilibrium could go undetected, but the conclusion does appear to be at least approximately valid.

Application of the foregoing argument yields an important relation between the radius of the pores and their centre-to-centre spacing L or their number per unit volume n. The solid is considered an assembly of cells, each of which has a pore at its centre. The number of gas atoms in any cell is $L^3 B/a^3$. If all the gas generated is in the pore, the pressure is given by

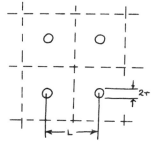

$$p = \frac{3Bk\Theta L^3}{4\pi a^3 r^3}.$$ (2–9)

Equating this pressure to the restraint due to surface tension, gives

$$r^2 = \left(\frac{3Bk\Theta}{8\pi\gamma a^3}\right) L^3 = \left(\frac{3Bk\Theta}{8\pi\gamma a^3}\right) \frac{1}{n}.$$ (2–10)

Thus, the pore density and radius are interrelated with the constant of proportionality depending on the operating temperature, the burnup, and properties of the material. It is worth emphasizing that the deduction is true only to the extent that the assumptions are valid; viz., the gas pressure is in equilibrium with the surface tension and a negligible amount of gas remains in the lattice. The analysis also assumes a regular array of uniform spherical pores, but, if applied with discretion, the relations of Eq. (2–10) can prove to be useful approximations.

With the same reservations, the macroscopic swelling can be calculated as

$$\frac{\Delta V}{V} = \frac{4\pi}{3}\left(\frac{r}{L}\right)^3 = \left(\frac{Bk\Theta}{2\gamma a^3}\right) r \approx \left(\frac{Bk\Theta L}{3\gamma a^3}\right)^{3/2} \approx \left(\frac{Bk\Theta}{3\gamma a^3}\right)^{3/2}\left(\frac{1}{n}\right)^{1/2}.$$

$$(2\text{–}11)$$

Thus, to keep the volume increases small, there should be a closely spaced distribution of fine pores. This expresses the physical facts that, if a large amount of gas is to be contained in a small volume, it must be at a high pressure and that high pressures can be balanced by surface tension only if the pores are small.

The belief that the fission-product gases have an extremely low solubility in uranium is usually based on calculations. Rimmer and Cottrell (1957) deduced a very high heat of solution for xenon in copper of five to ten electron volts, and application of their method to uranium gave comparable results. Blin (1958) arrived at substantially lower values, but they were still in excess of one electron volt so that the solubility calculated remained very low. The algebraic sign of the heat of solution is such that the solubility should decrease with decreasing temperature, from a low value even at high temperatures. Many experiments have shown negligible permeability of rare gases through metals. With the more recent demonstration that the same gases can migrate when they have been introduced into the metals, the inference is strong that their solubilities are extremely low. Specifically for uranium,

Johns (1944) reported that the solubility of argon at 600°C was not more than one part in 3×10^5. Although there is an undeniable need for direct experimental measurements of the solubility of xenon and krypton in several fuel materials, the assumption of low solubilities appears reasonable.

A consequence of the relatively large heat of solution estimated for the rare gases is that very few, e.g., four, gas atoms with associated vacancies could form a stable nucleus for a pore. Thus, the nucleation of the pores does not present a serious problem. If impurities or imperfections do not provide enough nuclei for heterogeneous nucleation, homogeneous nucleation could occur without difficulty. Greenwood and Speight (1963) made approximate estimates of the number of atomic jumps a gas atom would have to perform to reach an existing nucleus or to nucleate a fresh pore. Equating the two numbers gave the pore separation above which homogeneous nucleation should occur as

$$L^6_{\min} \approx \mu_0 D_g r^2 a^2 \left| \frac{dB}{dt} \right., \qquad (2\text{--}12)$$

where D_g is the diffusion coefficient for the gas and the numerical constant μ_0 probably lies in the range from 10 to 100. Greenwood and Speight reported that typical pores in their irradiated uranium had diameters about 0.1μm and a spacing about 0.6 μm However, Bainbridge and Hudson (1965), using an improved technique for replicating the same specimens, found the pores to be about 0.02 μm diam. with a spacing about 0.1 μm. These latter values are compatible with homogeneous nucleation (Eq. 2–12), but, since the observed spacing is close to that which might be expected for dislocations, heterogeneous nucleation on dislocations is also possible. Either way, a spacing of roughly 0.1 μm can be readily explained.

The easy nucleation of pores makes the measurement of diffusion coefficients for the gases difficult and unreliable. Values for the coefficients reported by different investigators, and even by the same investigator for different specimens, vary widely. At 700°C, for instance, values reviewed by Savage (1963), range from 10^{-11} down to 10^{-15} cm^2/sec. Since the lower values for the coefficient were obtained from specimens with high burnup where much of the gas would be in pores, the higher values are favoured. However, there is little confidence that even these represent true diffusion coefficients, since at a burnup as low as $10^{-5}\%$ the internally trapped gas exceeded that released.

As irradiation proceeds, the nuclei presumably capture further gas atoms and vacancies and develop into the pores that have been observed by electron and optical microscopy (Sec. 2.9). The relative number of gas atoms and vacancies entering a pore will be governed by the requirement to balance the gas pressure by the surface tension. However, many experiments summarized by Greenwood and Boltax (1962) showed that post-irradiation annealing caused not only an increase in the size of pores but also a decrease in their number. Since these changes are in the direction of greater swelling (Eq. 2–11) much consideration has been devoted to mechanisms whereby pores grow at the expense of their neighbours.

From analogy to the behaviour of solid precipitates in metals, it seemed possible that the gas atoms dissolved from the small pores, diffused through the solid and entered the large pores. The mechanism would be one of atomic diffusion down a concentration gradient. Greenwood and Boltax (1962) have deduced how the growth of a pore should depend on the heat of solution of the gas in the metal. If for every n_g gas atoms that enter a pore of radius r, n_v vacancies also enter, the rate of volume increase for the pore can be written

Flux of gas atoms $= -D_g \frac{\partial c}{\partial x}$

$$4\pi r^2 \frac{dr}{dt} = \left(\frac{n_v + n_g}{n_g}\right) D_g 4\pi x^2 \frac{\partial c}{\partial x}, \qquad (2\text{–}13)$$

where c is the gas concentration (expressed as an atomic fraction) at radius x from the centre of the pore. But the general gas equation gives

$$\frac{2\gamma}{r}(n_g + n_v)a^3 = n_g k\Theta. \qquad (2\text{-}14)$$

Hence,

$$r\frac{dr}{dt} = \left(\frac{k\Theta D_g}{2\gamma a^3}\right) x^2 \frac{\partial c}{\partial x}. \qquad (2\text{-}15)$$

Separating and integrating between the limits c_r and c_∞ yields

$$\frac{dr}{dt} = \frac{D_g k\Theta}{2\gamma a^3}(c_\infty - c_r). \qquad (2\text{-}16)$$

It thus becomes necessary to estimate how the gas concentration varies in the neighbourhood of the pore. To the extent that the gas solubility (expressed as volume gas per volume solid) is given by the approximation $\exp(-G/kT)$, where G is the free energy of solution of the gas, the concentration immediately outside the pore and in equilibrium with the pressure within can be written

$$c = \frac{pa^3}{kT}\exp(-G/k\Theta). \qquad (2\text{-}17)$$

If all the available gas is already in pores and only re-distribution between pores is being considered, the total number of gas atoms in all the pores remains constant, i.e.,

$$\Sigma\left(\frac{2\gamma}{r_i} \cdot \frac{4}{3}\pi r_i^3\right) = \text{constant}. \qquad (2\text{-}18)$$

Therefore,

$$\Sigma\left(r_i \frac{dr_i}{dt}\right) = 0, \qquad (2\text{-}19)$$

and from Eq. (2–16),

$$\Sigma r_i(c_\infty - c_i) = 0. \qquad (2\text{-}20)$$

If the pore of radius r is imagined as an unusually large one that grows at the expense of many surrounding pores whose mean radius r_m effectively determines the concentration c_∞ at a distance from the large pore as

$$c \approx \frac{2\gamma a^3}{r_m kT} \exp(-G/k\Theta), \qquad (2\text{-}21)$$

then

$$\frac{dr}{dt} \approx D_g \left(\frac{1}{r_m} - \frac{1}{r}\right) \exp(-G/k\Theta). \qquad (2\text{-}22)$$

Although a rigorous analysis requires knowledge of the initial value of r_m and how it varies with time, an approximate solution giving the functional relation is

$$r^2 \approx D_g t \exp(-G/k\Theta). \qquad (2\text{-}23)$$

Or, substituting

$$G = H - \Theta \Delta S_v \qquad (2\text{-}24)$$

and

$$D_g = (D_0)_g \exp(-Q_g/k\Theta), \qquad (2\text{-}25)$$

where H is the heat of solution of the gas, ΔS_v is the change in the entropy of lattice vibrations, and Q_g the activation energy for gas diffusion in the metal, gives

$$r^2 \approx (D_0)_g t \exp(\Delta S_v/k) \exp - (H + Q_g)/k\Theta. \qquad (2\text{-}26)$$

Even though the solubility of the gas in the metal lattice were too low to permit appreciable pore growth, passage of gas atoms from one pore to another through connecting dislocations might still be possible. However, transmission electron microscopy has shown pores situated on dislocations that ended on a free surface (Makin et al., 1962). If rapid passage through the dislocations were possible, it is unlikely that any pores would remain long enough to be seen. Barnes and Mazey (1963) have watched similarly situated pores of helium in copper and have seen them survive unchanged through several pulse anneals. They even observed pairs of pores of different sizes on the same dislocation maintain their original sizes. Their observations demonstrate that the solubility of helium in copper, either in the lattice proper or in dislocations, is negligibly small. Since krypton and xenon have much larger atoms than helium, it is most unlikely that re-solution of the fission-product gases plays an important role in pore growth, during post-irradiation annealing of uranium.

The same conclusion need not necessarily apply to pore growth during irradiation. The low solubility means that the rate-controlling stage is the entry of gas atoms into the metal surrounding the small pores. Any process that introduces gas atoms into the metal so that the resulting concentration in the metal increases with increasing concentration in the adjacent gas phase provides the necessary condition: Diffusion could then occur down the concentration gradient between pores of different sizes. During irradiation, the knock-on of gas atoms by fission fragments passing through the pore could yield such an effect (Lewis, 1960). If pore growth were to occur by this mechanism, the rate would not be expected to be that derived by Greenwood and Boltax for post-irradiation annealing (Eq. 2–26), since the heat of solution would not enter into the analysis.

The same work on helium in copper (Barnes and Mazey, 1963) demonstrated another mechanism for pore growth. Comparison of the same area of the thin film by electron microscopy after successive pulse anneals *in situ* revealed that individual pores had migrated at temperatures well below the melting point. The specimens were heated by removing the condenser aperture: In this state the microscope did not have good resolution, but the image was of sufficient quality to permit observation of pore movement during annealing. The general direction of movement of all pores was the same, presumably up the thermal gradient which was very severe in these specimens. As an approximation, the velocity of a pore was inversely proportional to its radius. Pores were seen to disappear both by reaching a free surface of the film and by coalescing with other pores.

The authors argued against sublimation being responsible for the movement in these specimens and concluded that surface diffusion of atoms around the periphery of the pores provided the mechanism. In a thermal or pressure gradient, all pores would move in the same direction with the smaller pores moving more rapidly than the larger ones. Thus, collisions should occur, causing pore growth. Even without an externally imposed gradient, collisions could still occur by random motion of the pores (very similar to Brownian motion of particles suspended in a liquid). Once proposed, the concept seems obvious. A single atom, or a vacancy, or the two together, can diffuse easily, whereas a macroscopic pore is virtually immobile. However, between these two extremes there is a whole range of atom-vacancy clusters, just-discernible pores, sub-microscopic and microscopic pores that are associated with a continuous decrease in mobility. From such a point of view, the distinction between dissolved and precipitated gas is arbitrary, so there may be no clear demarcation between a re-solution mechanism and one of pore migration.

When an atom on the surface of a pore jumps one atomic spacing a, with a frequency $v_D\psi \exp(-Q_s/k\Theta)$, where v_D is the Debye frequency, ψ is an entropy factor, and Q_s is the activation energy, the entire pore moves a distance proportionally less by the ratio of the atom volume to the pore volume, i.e., a distance $3a^4/4\pi r^3$.* Since there are $4\pi r^2/a^2$ atoms on the pore surface the jump frequency of the pore is $4\pi(r/a)^2 v_D\psi \exp(-Q_s/k\Theta)$. Hence the diffusion coefficient of the pore D_p is given by

$$D_p \approx \frac{3a^6}{4\pi r^4} v_D\psi \exp(-Q_s/k\Theta) = \frac{3}{8\pi}\left(\frac{a}{r}\right)^4 D_s, \quad (2\text{--}27)$$

where D_s is the coefficient for atomic surface diffusion. This expression reflects the greater mobility of smaller pores and predicts an exponential dependence on temperature. However, the fact that surface diffusion does not require the generation of a vacancy means that the

---

* The following derivations are synthesized from those of Greenwood and Speight (1963), Barnes (1964), Speight (1964 a & b), Shewmon (1964), and Gruber (1965) without exactly reproducing any of the sources.

variation of D_p with temperature should be less rapid than that of a bulk diffusion coefficient.

If the pore is in a thermal or pressure gradient, or if it is being dragged by a dislocation or a grain boundary, it will be acted on by a force F. The pore will, therefore, have a drift velocity v superimposed on its random motion. From the Nernst-Einstein relation,

$$v \approx \frac{D_p}{k\Theta} F \qquad (2\text{–}28)$$

and, hence,

$$v \approx \frac{3}{8\pi} \frac{D_s}{k\Theta} \left(\frac{a}{r}\right)^4 F. \qquad (2\text{–}29)$$

In a thermal gradient, the force on an individual atom f can be expressed as the enthalpy gradient or

$$f = - C_p \frac{d\Theta}{dx} \approx - 3k \frac{d\Theta}{dx}, \qquad (2\text{–}30)$$

where C_p, the specific heat at constant pressure, is taken approximately equal to $3k$. Regarding the motion of an atom in one direction as equivalent to that of a vacancy in the opposite direction, the force on the pore is obtained by multiplying $-f$ by the number of atom sites in the pore, i.e.,

$$F\text{(thermal gradient)} = \frac{4\pi k}{a^3} \frac{d\Theta}{dx} r^3. \qquad (2\text{–}31)$$

Consider a pore located on a dislocation. If the dislocation moves, because of either an imposed mechanical stress or climb in a nonequilibrium concentration of vacancies, the pore will be subjected to the forces of the dislocation line tension. In the limit that both arms of the dislocation pull in the same direction, the maximum force on the pore is given by

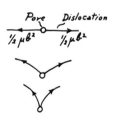

$$F\text{(dislocation)} = \mu b^2, \qquad (2\text{–}32)$$

where μ is the shear modulus of the material and b is the Burgers vector of the dislocation. For most metals, this force is of the order of 10^{-4} dyn. In contrast, a pore attached to a grain boundary will be acted on by a maximum force given by

$$F(\text{grain boundary}) = \pi\gamma_{gb}r, \qquad (2\text{--}33)$$

where γ_{gb} is the interfacial energy of the grain boundary.

If it is assumed that γ_{gb} has a value of 10^3 dyn/cm and that the pore radius must be greater than one angstrom, the force of a grain boundary on a pore is always greater than that of a dislocation, and the difference increases as the pore grows. Thus, grain boundaries are more effective than dislocations in trapping and sweeping pores, especially large ones. For comparison, in a thermal gradient of 100 deg C/cm, the force given by Eq. (2–31) equals that due to a dislocation for a pore radius 0.1 µm, but the thermal gradient force rapidly becomes the more important as the pore size increases.

The pore velocity is obtained by substituting for F in Eq. (2–29). Thus, in a thermal gradient

$$v \approx \frac{3aD_s}{2\Theta} \frac{d\Theta}{dx} \frac{1}{r}, \qquad (2\text{--}34)$$

i.e., the velocity should be inversely proportional to the pore radius, as found by Barnes and Mazey (1963) for their original experiments on pores in copper. A mechanism depending on the pore being dragged by dislocations or grain boundaries would predict the velocity to be proportional to some other power of the radius.

To deduce the pore growth to be expected as a result of random motion of the pores, the mean coalescence time t_c is calculated. The pore diffuses through a distance equal to its own diameter in a time $4r^2/6D_p$, i.e., in that time it has moved into an adjacent pore *site*. However, if L is the pore spacing, $(L/2r)^3$ sites are associated with each pore. Thus, the probability of coalescence with another pore in making the unit move is $Z(2r/L)^3$, where

Z is the number of new neighbouring sites explored by the pore. Therefore,

$$t_c \approx \frac{4r^2}{6D_p}\left(\frac{L}{2r}\right)^3 \frac{1}{Z} = \frac{L^3}{12ZD_pr}. \qquad (2\text{-}35)$$

By substituting for D_p from Eq. (2–27) and for L^3 from Eq. (2–10),

$$t_c \approx \frac{2\gamma}{aBk\Theta D_s}r^5. \qquad (2\text{-}36)$$

In an idealized process, all the pores are of radius r_0 at zero time and coalesce in pairs after a time t_c to form half as many pores of radius r_1. From Eq. (2–18)

$$r_1^2 = 2r_0^2. \qquad (2\text{-}37)$$

Since t_c varies as r^5, it takes $2^{5/2}$ times as long for the next set of collisions, and so on. Thus, for m sets of collisions, during which the pores have grown from a radius of r_0 to one of $r_0(\sqrt{2})^m$, the time taken is

$$t = t_c[1 + 2^{5/2} + (2^{5/2})^2 + \cdots + (2^{5/2})^{m-1}]$$

$$= \frac{t_c}{4.7}[(2^{5/2})^m - 1]. \qquad (2\text{-}38)$$

Therefore,

$$t \approx t_c(r^5 - r_0^5)/4.7r_0^5, \qquad (2\text{-}39)$$

and use of Eq. (2–36) to solve for r, gives

$$r \approx \left(\frac{2aBk\Theta D_s t}{\gamma} + r_0^5\right)^{1/5}. \qquad (2\text{-}40)$$

If the spacing of dislocation lines is sufficiently small, pores may nucleate on the dislocations or may be trapped on them after an initial period of random migration. The pores are then constrained to move along the dislocations by a series of random jumps in either of only two directions. If the spacing of dislocation lines is λ_d, then the spacing of pores along a line λ_p can be obtained from the general gas equation. Thus,

$$\frac{2\gamma}{r} \frac{4\pi r^3}{3} \frac{1}{\lambda_p \lambda_d^2} = \frac{Bk\Theta}{a^3} \qquad (2\text{-}41)$$

and

$$\lambda_p = \frac{8\pi\gamma a^3 r^2}{3Bk\Theta\lambda_d^2}. \qquad (2\text{-}42)$$

The mean coalescence time for these pores is, therefore,

$$t_c \approx \lambda_p^2/2D_p \qquad (2\text{-}43)$$

$$\approx \frac{250\gamma^2 a^2}{B^2 k^2 \Theta^2 \lambda_d^4 D_s} r^8.$$

The use of the same method as was employed for truly random motion yields a relation for the pore growth

$$r \approx \left(\frac{B^2 k^2 \Theta^2 \lambda_d^4 D_s}{16\gamma^2 a^2} t + r_0^8\right)^{1/8}. \qquad (2\text{-}44)$$

Pore coalescence and, hence, swelling can occur when the pores are acted on by directed forces as well as when they migrate randomly. However, it is more difficult to construct a credible model that is susceptible to quantitative analysis. Barnes (1964) considered an array of pores all the same size and all sunject to the same force F: He derived an expression for pore growth during post-irradiation annealing as

$$r \approx \left(\frac{3aBFD_s t}{16\pi\gamma}\right)^{1/4}, \qquad (2\text{-}45)$$

and during irradiation at temperature as

$$r \approx \left[\frac{3a\dfrac{dB}{dt}FD_s t^2 \log_e t}{16\pi\gamma}\right]^{1/4}. \qquad (2\text{-}46)$$

If the forces on the pores result from an energy gradient or movement of dislocations, all the pores would be driven in the same direction, but, since they are all assumed to be the same size, they would all move with the same velocity. Collisions are possible, therefore, only if each pore moves intermittently while the remainder are at rest. Such a condition might be satisfied for certain forms of dislocation movement, but the model does not appear to have wide application.

Gruber (1965) considered pores driven by forces that all acted in the same direction, but, by postulating two groups of pores of different radii, he could obtain collisions if the pore velocity was radius-dependent. For a force given by

$$F = Cr^q, \tag{2-47}$$

where C and q are constants, the pore velocity can be derived from Eq. (2-29) as

$$v \approx \frac{3}{8\pi} C \frac{D_s a^4}{k\Theta} r^{q-4}. \tag{2-48}$$

For equal numbers of pores of radius r_1 and r_2, where

$$r_1 = (1 - c)\bar{r} \quad \text{and} \quad r_2 = (1 + c)\bar{r}, \tag{2-49}$$

a given pore of one size collides in a time Δt with all pores of the other size whose centres lie within a cylinder of radius $(r_1 + r_2)$ and length $(v_1 - v_2)\Delta t$. Since there are $n/2$ pores of each size per unit volume, the number of collisions per unit volume is given by

$$\Delta n = (r_1 + r_2)^2 (v_1 - v_2)\Delta t \left(\frac{n}{2}\right)^2 \tag{2-50}$$

$$\approx \frac{3}{8} \frac{C D_s a^4 n^2}{k\Theta} [(1 - c)^{q-4} - (1 + c)^{q-4}] r^{q-2} \Delta t.$$

Application of Eqs. (2-18) and (2-49) shows that, as a result of the collisions, the mean radius of the pores has changed by an amount

$$\Delta \bar{r} = [\sqrt{2}(1 + c^2)^{\frac{1}{3}} - 1] \frac{\Delta n}{n} \bar{r} \tag{2-51}$$

Therefore, substituting from Eq. (2–50) and using the general gas equation in the form

$$\frac{2\gamma}{r} n \frac{4\pi r^3}{3} = \frac{B}{a^3} k\Theta ,$$ (2–52)

gives

$$\Delta\bar{r} = [\sqrt{2}(1 + c^2)^{\frac{1}{2}} - 1][(1 - c)^{q-4} - (1 + c)^{q-4}]$$

(2–53)

$$\times \frac{9}{64\pi} \frac{BCD_s a}{\gamma} \bar{r}^{q-3} \Delta t .$$

For the particular case of motion in a thermal gradient, values of C and q are obtained by comparing Eqs. (2–31) and (2–47). Substituting in Eq. (2–53) and passing to the limit yields

$$d\bar{r} = [\sqrt{2}(1 + c^2)^{\frac{1}{2}} - 1] \left[\frac{c}{1 - c^2}\right] \frac{9BkD_s}{8a^2\gamma} \frac{d\Theta}{dx} dt .$$ (2–54)

By integration,

$$\bar{r} = fn(c) \frac{9BkD_s}{8a^2\gamma} \frac{d\Theta}{dx} t ,$$ (2–55)

where the initial value of $\bar{r}$ has been ignored in comparison to the final one. Obviously the assumption of only two sizes of pores is unrealistic. However, it might be expected that, if the analysis could be performed for a more representative pore-size distribution, only the factor $fn(c)$ would be affected, leaving

$$\bar{r} = A_0 \frac{BD_s}{a^2\gamma} \frac{d\Theta}{dx} t ,$$ (2–56)

where A_0 is characteristic of the initial pore-size distribution.

Where the pores are all of the same radius, Eq. (2–11) can be used to derive the swelling from the relation for the pore growth. However, where a distribution of pore sizes exists, it would be incorrect to use the mean radius to calculate the swelling in this simple fashion, since

$$\sum_{i=1}^{N} \left(n_i \frac{4}{3} \pi r_i^3 \right) \neq N \frac{4}{3} \pi \left[\frac{\sum_{i=1}^{N} (n_i r_i)}{\sum_{i=1}^{N} (n_i)} \right]^3. \qquad (2\text{--}57)$$

To the extent that a distribution of pore sizes always exists in practice, this step must be regarded as an approximation in all the analyses that assume a uniform radius to simplify the calculations. Comparison is made therefore between the various mechanisms in Table 2–1 by contrasting their predictions for the pore growth rather than for swelling.

Thus, all the mechanisms analysed lead to a relation of the same general form: The pore radius increases exponentially (or nearly so) with increasing temperature, linearly with some power of the annealing time, and linearly with some power of the burnup (except for the gas solubility mechanism in which the analysis does not determine the dependence on burnup). Ideally, an identification of the mechanism could be made from experimental results by determining the values of the activation energy and powers concerned. Since surface diffusion, which does not require the formation of a vacancy, commonly has an activation energy roughly half that for bulk diffusion, the gas-solubility mechanism probably has the greatest temperature coefficient. At the other extreme, pore migration along dislocations should be least sensitive to temperature. The latter mechanism should also become the slowest rate of pore growth ($r \propto t^{1/8}$), while migration of pores in a thermal gradient should become the most rapid ($r \propto t$). In practice, such criteria have not yet served to identify positively any mechanism.

When uranium that had been irradiated at low temperatures was subsequently annealed at 575°C, the swelling varied as $t^{1/4}$ (Churchman et al., 1958). Provided that the pores could be considered of uniform size, this result is in exact agreement with one of the pore-migration mechanisms. However, it is probably also consistent with the relation ($r \propto t^{1/5}$) deduced for random motion of pores. Rich and Barnes (1959) annealed irradiated uranium for one hour at various temperatures from 300 to 950°C

Table 2-1
Post-Irradiation Pore Growth Predicted by Various Mechanisms

Mechanism	Constant	Temperature	Time	Burnup
Gas solubility	$r \approx \{(D_0)g \exp(\Delta S_v/k)\}^{1/2} \times$	$\{\exp - (H + Q_g)/k\Theta\}^{1/2} \times$	$t^{1/2}$	
Random pore migration	$r \approx \left\{\dfrac{4a^3 k v_D \psi}{\gamma}\right\}^{1/5} \times$	$\{\Theta \exp - Q_s/k\Theta\}^{1/5}$	$\times \; t^{1/5}$	$\times \; B^{1/5}$
Pore migration along dislocations	$r \approx \left\{\dfrac{k^2 \lambda_d^4 v_D \psi}{8\gamma^2}\right\}^{1/8} \times$	$\{\Theta^2 \exp - Q_s/k\Theta\}^{1/8}$	$\times \; t^{1/8}$	$\times \; B^{1/4}$
Dislocations dragging pores	$r \approx \left\{\dfrac{3a^3 F v_D \psi}{8\pi\gamma}\right\}^{1/4} \times$	$\{\exp - Q_s/k\Theta\}^{1/4}$	$\times \; t^{1/4}$	$\times \; B^{1/4}$
Pores in thermal gradient	$r \approx \left\{\dfrac{2A_0 v_D \psi}{\gamma}\right\} \times$	$\left\{\dfrac{d\Theta}{dx} \exp - Q_s/k\Theta\right\}$	$\times \; t$	$\times \; B$

and presented their results in the form of a graph of log n versus $1/\Theta$. Apart from the values at the highest temperature, the experimental points lay on a straight line with a slope 0.15 eV. If the same assumption of pore-size uniformity is made, n varies as $1/r^2$ (Eq. 2–11). The temperature dependence of pore growth was therefore

$$r \propto \exp(-0.15\text{eV}/2k\Theta). \qquad (2\text{–}58)$$

Such a dependence is in agreement with expectations for a mechanism controlled by surface diffusion with its low activation energy, but a lack of experimental values for that activation energy prevents the alternative mechanisms being differentiated.

Greenwood and Boltax (1962) assembled the results from three independent experiments of annealing irradiated uranium in the gamma phase. For the same amount of swelling in each, they found the annealing time varied with an activation energy of 4.3 eV. It is, of course, dangerous to draw conclusions from only three points, each obtained from different material irradiated under different conditions. However, if the value is valid, it is probably inconsistent with a surface-diffusion mechanism and tends to support gas solubility. Adda et al. (1964) found the activation energy for diffusion of xenon in unirradiated uranium to be 88 kcal/g atom (3.83 eV) in the gamma phase and only 27 kcal/g atom (1.18 eV) in the beta phase. The former value is in fair accord with the swelling results and supports the interpretation of gas solubility in the gamma phase, while the latter low value suggests a grain boundary or surface-diffusion mechanism in the beta phase. Adda and his colleagues deliberately used very low xenon concentrations to avoid gas trapping in pores, but they introduced no proof that they had been successful. It is therefore possible that pore migration controlled the release at all temperatures: At sufficiently high temperatures, vacancy diffusion through the lattice, with its relatively high activation energy, may become more important than surface diffusion for pore movement.

Loomis and Pracht (1963) conducted extensive and self-consistent studies of the swelling of high-purity uranium during post-irradiation annealing. In the alpha phase, they found appreciable swelling at temperatures over 500°C, with an apparent activation energy of approximately 2.5 eV for the volume change at a particular annealing time: The swelling varied as the time raised to some power between $\frac{1}{4}$ and $\frac{1}{2}$ (Fig. 2–11). Microscopy revealed that much of

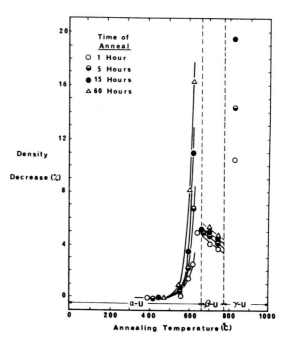

Fig. 2–11. Effect of the annealing temperature on the swelling of uranium previously irradiated below 300°C to a burnup between 0.24 and 0.30%. (After Loomis & Pracht (1963), courtesy *J. Nucl. Mater.*)

the volume change was due to large pores on grain boundaries in recrystallized regions, and that elsewhere the pores were smaller and more uniform. Thus, the dependence on time and temperature are probably representative of recrystallization, while the nonuniformity of pore size renders dubious any deduction of the pore-growth mechanism from the observed activation energy. Significantly,

in these post-irradiation studies, the authors did not observe the exaggerated swelling around 450°C, attributed
to in-reactor growth stresses.

From their observations, Loomis and Pracht concluded
that the grain boundaries were sweeping the pores along
during recrystallization. In other irradiated specimens,
however, recrystallization occurred without pores collect-
ing at the boundaries (Eldred, 1961). Speight and Greenwood (1964), in a theoretical analysis, showed that the
pores should be swept along only if the difference in free
energy between adjacent grains that causes boundary
movement is not too large. They proposed that this requirement could reconcile the conflicting observations.

Following annealing in the beta phase, Loomis and
Pracht observed a uniform distribution of porosity and
some cracking, mostly intergranular. The volume change
appeared to be largely due to the alpha-phase annealing
received by the specimen on being brought up to temperature. Thus, prolonged annealing in the beta phase
had little further effect on either the swelling or the pore
size. After annealing in the gamma phase, the volume
changes were larger than for the same time in the alpha
phase, but they were not as large as would have been
obtained by extrapolating from the lower temperatures.
Again, the porosity was heterogeneous with large pores
along the grain boundaries, sometimes elongated and
linked. The association with grain boundaries may be
due to these acting as sources of vacancies. If so,
the post-irradiation annealing results would not be
representative of in-reactor behaviour since irradiation
damage would be expected to provide a copious supply
of vacancies throughout the metal. Metallography of other
specimens irradiated at temperatures in the gamma phase
showed no especially large pores at the grain boundaries
(Greenwood, 1962).

There have been very large variations in the magnitude
of the swelling at a given temperature of post-irradiation
anneal: In this respect, at any rate, there is a strong resemblance to swelling during irradiation. For instance,

Churchman and Barnes (1958) measured a density decrease of 2% on annealing at 575°C uranium irradiated to 0.4% burnup below 300°C, while Eldred (1958) obtained an order of magnitude less swelling on annealing at the same temperature uranium irradiated to 0.17% burnup at about 300°C. Part of the difference may well be caused by differences in burnup and temperature of irradiation, as these may affect the scale of pore nucleation, and by differences in original grain size. However, Eldred's material had a much higher impurity content than that used by Churchman and Barnes, and another experiment by Loomis and Pracht has demonstrated the possible importance of impurities. A specimen containing 1100 parts per million (ppm) carbon showed less swelling on annealing at 620°C than did others containing only 17 ppm carbon. The authors suggested that the effect was due to the suppression of recrystallization known to result from a higher carbon content. When it is recalled that production uranium at the time of some of the earlier experiments commonly contained five volume percent of second-phase impurities, the large variation in behaviour becomes understandable.

Once one of the pores becomes sufficiently large, it can continue to grow by vacancy diffusion alone. If such a pore, surrounded by many smaller ones of radius r and spacing L, increases its radius r_a by an amount Δr_a, it captures $4\pi r_a^2 \Delta r_a / L^3$ small pores.* Since each small pore contains $(4\pi r^3/3)(2\gamma/rk\Theta)$ gas atoms, the large pore captures $32\pi^2 r_a^2 \Delta r_a \gamma r^2 / 3k\Theta L^3$ gas atoms. Growth of the large pore will therefore be self-sustaining, if this number exceeds that needed for the extra pore volume, viz., $(4\pi r_a^2 \Delta r_a)(2\gamma/r_a k\Theta)$. Thus, the condition for this form of breakaway swelling is

$$r_a > \frac{3}{4\pi} \frac{L^3}{r^2}. \tag{2–59}$$

---

* This analysis, based on one by Greenwood and Boltax (1962), is due to Lewis (1965b).

Even when the pore sizes remain uniform, application of a similar argument indicates that self-sustaining growth of the pores should occur for a radius greater than that given by

$$r^3 = L^3/4\pi \qquad (2\text{-}60)$$

(Barnes, 1964). The use of Eq. (2–11) shows that this stage will be reached at a swelling of

$$\frac{\Delta V}{V} = \frac{1}{3}. \qquad (2\text{-}61)$$

For a cubic array of uniform pores with each touching its nearest neighbours, the volume ratio of void-to-solid is $\pi:(6 - \pi)$. Thus, for a swelling of about 100%, the porosity should be largely interconnected, and the gas virtually all released.

Pores confined to the grain boundaries should become interconnected by the time

$$\frac{\Delta S}{S} = \frac{\pi}{4}, \qquad (2\text{-}62)$$

where $\Delta S/S$ is the fraction of the grain-boundary surface covered by pores. If in unit volume of fuel there are n_b grain-boundary pores, each of radius r_b, on grains of diameter d, then

$$\frac{\Delta S}{S} = \frac{r_b^2 n_b d}{6}. \qquad (2\text{-}63)$$

Where the pores trapped on the boundary contain all the gas atoms from a thin layer of thickness δ adjacent to the boundary, the general gas equation gives

$$\frac{2\gamma}{r_b} n_b \frac{4\pi r_b^3}{3} = \frac{6\delta}{d} \frac{Bk\Theta}{a^3}. \qquad (2\text{-}64)$$

Since

$$B = yz, \qquad (2\text{--}65)$$

where y is the fractional fission yield for rare gases, and z is the fractional burnup by fission of uranium atoms, substitution in Eq. (2–63) shows that by a burnup given by

$$z = \frac{2\pi y a^3}{3 y \delta k \Theta} \qquad (2\text{--}66)$$

the grain-boundary pores should become interconnected and release their gas. At this stage, the swelling due to grain-boundary porosity alone should be

$$\left(\frac{\Delta V}{V}\right)_b = \frac{4\pi r_b^3 n_b}{3} = \frac{2\pi r_b}{d}. \qquad (2\text{--}67)$$

Extensive grain-boundary porosity has been observed in UO_2 (Fig. 3–19) and UC (Fig. 4–1d).

In summary, therefore, a decade of research has shown that the expression "swelling" encompasses several phenomena. In addition to the inevitable volume increase due to solid fission products, the minimum swelling occurs for a distribution of uniform small pores. Even in this idealized situation pores progressively coarsen, probably by migration of individual pores either randomly or under the action of one or more of the forces discussed. However, opposed irradiation growth in adjacent grains of alpha uranium produces cracks and large pores at the boundaries and, hence, a maximum in the swelling around 450°C. Recognition of this fact should finally dispose of the widely held belief that in-reactor swelling at a given temperature can be accurately reproduced by post-irradiation annealing at that temperature. Recrystallization of alpha uranium can also result in exaggerated swelling. Once a few of the pores reach a critical size, their growth is accelerated by the capture of surrounding smaller pores. Even with a uniform pore distribution, breakaway swelling is to be expected at a volume increase of approximately 30%.

Naturally, much of the work in this field has had as its aim the achievement of low values for swelling. One of several theories tested was that uranium irradiated first at low temperatures, to provide a very fine dispersion of pores, would be more resistant than fresh material to swelling during subsequent service at elevated temperatures. Greenwood (1961) examined a uranium bar that had experienced, albeit by accident, just such a sequence with a continuous range of elevated temperatures along the bar during the second period. He found that the pores did not maintain their spacing: Rather, the number decreased, and their size increased as the temperature increased. These observations constitute good evidence for the growth of large pores at the expense of small ones by one or other of the mechanisms already discussed.

The only attempts to reduce swelling that have proved successful have involved alloying the uranium, and these will be considered in Sec. 2.7. First, however, it is necessary to examine the effect of irradiation on phase transformations.

2.6 Phase Transformations under Irradiation

Since the local temperatures in fission spikes are very high, but return to ambient extremely rapidly, irradiation might be expected to have the same effect on phase transformations as a quench from a high temperature. Although some such effects have been observed, the analogy is not exact. The difference is largely due to the small volume and short duration of the spike which suppress the nucleation of phases not already present. The phenomena principally observed are, therefore, ordering, disordering, homogenization, and alteration in the proportion of two phases coexisting.

Tucker and Senio (1956) demonstrated experimentally the difference between irradiation and quenching by taking advantage of the fact that chromium additions to uranium permit the retention of the beta phase to room temperature. A specimen of U–2at.%Cr, quenched to retain the beta structure, was irradiated with another of the same composition that had been annealed to contain only the equilibrium alpha phase. After irradiation at a temperature well below the alpha/beta transformation, both were examined by x-ray diffraction. Up to 2×10^{17} fissions/cm^3, neither had changed: The alpha phase specimen exhibited only the alpha structure and the beta phase specimen exhibited only the beta structure. Thus, the disrupted material in the fission spike had grown back on the crystal lattice of the surrounding matrix, regardless of whether or not it was the equilibrium phase. Where two phases are present, fission spikes straddling a phase boundary might conform more rapidly to one lattice than to the other. Each spike would thereby transform a fraction of its volume, and eventually the whole specimen could be converted to a single-phase structure.

Laboratory studies have shown that the beta- to alpha-phase transformation in uranium occurs only after a finite incubation period that depends on the material's purity. A specimen irradiated in the beta phase and then cooled below the transformation temperature while still under irradiation would retain its beta structure for a time equal to the incubation period: This can be several minutes. If in that time every atom is involved in a fission spike, the beta phase will be retained indefinitely. Beta-phase retention to lower temperatures should therefore be favoured by high fissioning rates and a sluggish transformation, in a manner derived quantitatively by Kehoe (1958), but no experimental test of this suggestion has been reported.

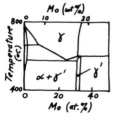

Many of the studies of phase transformations under irradiation have been concerned with uranium-molybdenum alloys. At about 16 wt% Mo (32 at.%), the phase in equilibrium at temperatures below 575°C is termed gamma-prime, since it is a tetragonal ordered form of the gamma-uranium structure. Bloch (1960) irradiated such specimens at a temperature under 65°C and followed the structural changes as a function of exposure by post-irradiation x-ray diffraction. She found that the ordering of the gamma-prime structure decreased continuously and had effectively disappeared by 10^{17} fissions/cm^3. Simultaneously, the cell parameters a' and $c'/3$ tended to the value of a for the cubic gamma structure. (The height of the gamma-prime unit cell is approximately three times that of the gamma unit cell.) The most notable observation was that the tetragonal phase transformed continuously into the cubic one without the two phases coexisting at any time.

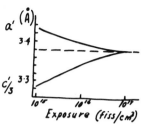

Exposure (fiss/cm³)

Previously, Konobeevsky et al. (1955) had reported on the behaviour of U–9wt%Mo (20 at.%Mo) irradiated to 10^{19} fissions/cm^3 at a temperature around 50°C. By comparing pre- and post-irradiation measurements of density, crystal structure, electrical resistivity, and the thermal coefficient of resistivity, they showed that a gamma structure retained by quenching before irradiation remained unchanged. In contrast, a specimen of the same composition that had been annealed to produce the alpha plus gamma-prime structure stable at low temperatures transformed to the gamma structure during irradiation.

By measuring the electrical resistivity after irradiation, Bleiberg et al. (1956) obtained the same result for alloys containing 9 to 13.5wt%Mo. The resistivity of U–9wt%Mo was measured during irradiation at temperatures down to −90°C in a later investigation (Bleiberg, 1959). The gamma-quenched specimen showed an initial decrease in resistivity prior to 10^{17} fissions /cm^3, corresponding probably to a small amount of ordering in the cubic structure. In the same initial period, the alpha plus gamma-prime specimens showed an appreciable increase, due presumably to the disordering of the gamma-prime phase, as already discussed. At higher exposures, the gamma specimen maintained an essentially constant resistivity, while the resistivities of the others continued to increase, but at a diminishing rate. Thus, these were approaching the value for the gamma-phase specimen, and by 10^{20} fissions/cm^3

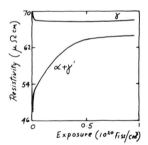

Exposure (10^{20} fiss/cm³)

they were within 7% (cf approximately 25% for the un-irradiated specimens).

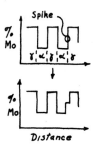

The microstructure of the annealed specimens consists of nearly pure alpha-uranium lamellae in a matrix of the gamma-prime phase. The lamellar spacing is typically 0.3μm. Konobeevsky and Bleiberg, with their respective colleagues, both supposed that an individual fission spike would even out the molybdenum concentration gradient within its volume, thus tending to homogenize the whole specimen. Concurrently, the gamma matrix (disordered gamma-prime phase) would grow at the expense of the alpha structure. Both groups independently derived an equation providing the concentration of molybdenum as a function of position and time in terms of the initial inter-lamellar spacing and a diffusion coefficient for molyb-denum. Using experimental results from the resistivity measurements, Bleiberg (1959) solved for this diffusion coefficient. He obtained a value independent of tempera-ture in the range from -90 to $+200°C$ and of neutron flux from 2×10^{12} to 2×10^{14} n/(cm^{2}sec), but directly proportional to the fission rate.

The good agreement obtained between the calculated and the observed kinetics of the resistivity changes for different interlamellar spacings argued in favour of the assumptions made in the derivation. However, Konobeev-sky et al. had supposed that homogenization resulted from thermally activated diffusion in the high-temperature region of the fission spike, whereas Bleiberg et al. had attributed the effect to displacement of atoms from the track of an energetic particle followed by collapse of the resulting shell. To resolve the question, Konobeevsky et al. (1962) irradiated Cu–8at.%Sn in a fast-neutron flux and Cu–8at.%Sn-1at.%Pu in a thermal-neutron flux. X-ray diffraction patterns showed that epsilon-phase particles (probably Cu_3Sn) were dissolved during irradiation of the latter but not the former, even when the number of dis-placed atoms should have been the same. Thus, the thermal effects of the fission spike appear to be more important than the displacement ones. Observations by Berman

(1965) on the kinetics of the phase transformation in ZrO_2–UO_2 during irradiation yield the same conclusion in another material.

In uranium-niobium alloys, too, irradiation-induced transformation to the gamma phase has been observed (Bleiberg et al., 1956). Specimens containing 10wt%Nb (22 at. %) were annealed for prolonged periods below 650°C to produce alpha-uranium lamellae in a matrix of the gamma-2 phase. Although irradiation below 200°C converted these two-phase specimens to the gamma phase stable at high temperatures, the rate of homogenization was much slower than in the uranium-molybdenum alloys. This observation was interpreted in terms of the greater tendency in the niobium alloys for like atoms to cluster together, as indicated by the existence of the ordered gamma-prime structure in the uranium-molybdenum system.

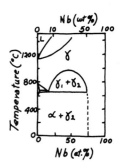

Studying swelling in uranium of 0.3% burnup during post-irradiation annealing, Loomis and Pracht (Sec. 2.5) found low values characteristic of the beta phase for an annealing temperature of 651°C. Since the alpha-beta transformation normally occurs at 660°C (or slightly higher on heating), it is possible that irradiation and the resulting fission products had lowered the transformation temperature and the point merits further investigation.

Irradiation can also affect the occurrence of precipitates. Konobeevsky et al. (1955) presented micrographs after successive periods of irradiation showing progressive breakup of cubic uranium-carbide precipitates in uranium metal. In contrast, Kramer et al. (1965) reported that irradiation at 300°C of U-4wt%Mo-0.1wt%Si alloys greatly increased the number of silicon-containing particles precipitated. Thus, fission spikes can both dissolve large particles and, by enhancing diffusion rates, accelerate the growth of very fine particles.

2.7 Effect of Alloying Additions

In examining claims that a certain alloying addition has beneficial effects on the irradiation behaviour of uranium, it is necessary to have a clear understanding of which aspects of the behaviour are being considered. Thus, many of the early proposals to add small amounts of chromium, zirconium, carbon, iron, aluminum, or silicon resulted from less growth or surface roughening having been observed in these alloys than in unalloyed uranium. The basic effect of these additions was to alter the transformation kinetics of the uranium so that a uniform, fine-grained, randomly oriented structure could be obtained by an acceptable heat treatment. Some experimental results by Zaimovsky et al. (1958) show clearly how the grain size of beta-quenched uranium is very sensitive to small amounts of iron, silicon, or aluminum, but how the effect saturates at a few hundred parts per million of the addition.

A second reason for alloying is, therefore, to give different batches of production uranium a reproducible reaction to heat treatment. The levels of impurity for iron, silicon, or aluminum in production uranium are normally in the range where small variations have a large effect on the grain size. By deliberate "doping", or "adjusting", all batches to the saturation level, the same variation in impurity content will have a negligible effect on the final grain size.

Since most of the early irradiations were conducted at relatively low temperatures, it is doubtful that the dimensional stability obtained provided any evidence for these alloying additions conferring resistance to swelling. On the contrary, a general conclusion from a search for swelling resistant, uranium-rich alloys was that most were no better than unalloyed uranium, while some were appreciably worse. The inferior behaviour may sometimes be attributed to compositional inhomogeneities. These would not only produce high internal stresses but would also result in localized regions of high burnup in which breakaway swelling might be expected to initiate.

Uranium-molybdenum alloys have, for several years, provided an exception to the generalization that no minor alloying addition is beneficial against swelling. Ball (1956) first reported decreased swelling in U-12 wt% Mo (25 at.% Mo) alloys. More recently, some well-controlled irradiations by Johnson and Holland (1964) have determined the effect of varying the amount of molybdenum for a few specific conditions. Increasing amounts of molybdenum, up to 11 wt%, reduced the swelling of specimens exposed to 3000 MWd/tonne U (0.3% burnup of the uranium) at about 675°C, 7000 MWd/tonne U (0.75%) at about 500°C, or 10 000 MWd/tonne U (1.1%) at about 670° C, as illustrated in Fig. 2–12. Since the curves are continuous

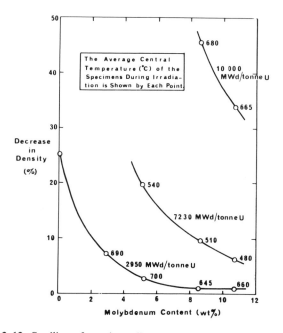

Fig. 2–12. Swelling of uranium alloys as a function of molybdenum content for different values of burnup and fuel temperature. (After Johnson & Holland (1964).)

and smooth, there is apparently no minimum amount of molybdenum required to give a benefit: Even three or four weight percent has had a large effect, particularly at the lower burnup.

Generally similar results had been obtained by Leeser et al. (1958) from the irradiation of 9 to 11 wt%Mo alloys. Those exposed to less than 2.0% burnup of all atoms had density decreases of under 5% for irradiation temperatures under 600°C but up to 25% between 600 and 800°C. Specimens of higher burnup experienced similar large swelling at temperatures of 300 to 400°C. These authors also conducted a limited investigation of the effects of fabrication method (rolling and extrusion) and heat treatment (gamma-quenched and alpha-plus-gamma-prime-transformed), but no significant difference in swelling behaviour was observed.

Subsequent experiments by Shoudy et al. (1962) demonstrated that in U-10wt%Mo alloys the swelling depended on the rate of fissioning as well as on the temperature and burnup: Leeser et al. had been fortunate in using high fissioning rates. The high-temperature gamma phase is normally retained to room temperature by quenching these alloys containing around 10wt%Mo, but annealing below the transformation temperature, e.g., in the range 500 to 565°C, causes the gamma phase to decompose into an $(\alpha + \gamma')$-phase mixture, where gamma-prime is a tetragonal modification of the cubic gamma phase. As already discussed (Sec. 2.6), irradiation effectively opposes the thermal annealing, so that for each temperature there is a critical fissioning rate above which the gamma phase is retained. Experimental results summarized by Kittel et al. (1964) show that conditions favouring retention of the gamma phase markedly reduce swelling (Fig. 2–13).

Originally, molybdenum alloys were investigated because the gamma-phase retention would eliminate forms of damage associated with the anisotropic alpha structure and with thermal cycling through the transformations. The increased strength of these alloys was also expected to reduce their swelling, but now it appears doubtful that the metal's strength seriously affects the swelling until the volume increase has already become excessive for most purposes. Measurements by Adda et al. (1964) suggest that an addition of one weight percent molybdenum to uranium decreases the diffusion rates for xenon by two orders of magnitude at 720 and 920°C. Even if true atomic diffusion was not occurring (Sec. 2.5), the results are still

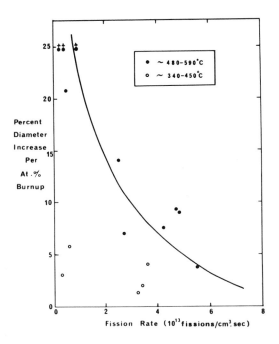

Fig. 2–13. Effect of the fission rate on the swelling of U–10wt%Mo
rods. (After Kittel et al. (1964), courtesy United Nations.)

relevant to swelling and can be interpreted as showing that
molybdenum somehow reduces the mobility of xenon in
uranium.

At low molybdenum contents, say two to five weight
percent, the situation is complicated, partly because of
the variety of structures obtainable by alteration of the
heat treatment. Relevant measurements on swelling of
these compositions have been made by Bentle (1961),
Mustelier et al. (1961), and Smith (1957). In general,
specimens that had been quenched from the gamma phase
exhibited much more swelling (15 to 30% volume increase
at 0.5% burnup of all atoms) than did those that had been
slow-cooled (3% increase at the same burnup) for irradiation temperatures around 500°C. Johnson (1962), drawing
on an interpretation of laboratory creep experiments,
suggested an explanation of the swelling observations: The

slow-cooling produced a structure composed of alpha-phase lamellae in a gamma-phase matrix, while the quench produced gamma-phase lamellae in alpha-phase grains. Thus, the material from the quench heat treatment was susceptible to much of the irradiation damage specific to alpha uranium, especially the enhanced grain-boundary swelling around 450°C.

Barnes et al. (1964) and Pugh and Butcher (1964) reported unpublished work by Bellamy on U-2.5wt% Mo specimens. Despite four very different heat treatments, all showed volume increases of only 5% after irradiation to approximately 0.5% burnup of all atoms at 560 to 585°C. However, it now appears (Sec. 2.5) that even unalloyed uranium, irradiated in this temperature range just above the limit for irradiation growth, might not exhibit large swelling. Thus, the information available is insufficient to establish that the molybdenum addition was beneficial, or that the heat treatment is unimportant. Pugh and Butcher also considered in some detail possible metallurgical changes in all these uranium-rich alloys, and emphasized that many of the observed differences in swelling between specimens may be due to small variations in their content of in-soluble impurities.

Alloys containing under 2 wt%Mo have been used extensively as fuel for French reactors (Englander et al., 1964): U-0.5 wt% Mo achieved 7500 MWd/tonne at 250 to 370°C, U-1.1wt% Mo achieved 10 500 MWd/tonne at 260 to 400°C, and U-1.5 wt% Mo achieved 7100 MWd/tonne at 260 to 400°C without any troublesome deformation or swelling. However, both the 0.5 and 1.1 wt% Mo alloys, when irradiated around 420 to 450°C, exhibited the same exaggerated swelling seen in unalloyed uranium (Fig. 2–8). In post-irradiation annealing of specimens irradiated at low temperatures, Englander et al. found little swelling below 600°C but large amounts over 700°C. Naturally, the exaggerated swelling associated with irradiation growth was not apparent in these tests.

Uranium-niobium alloys, containing 9 to 12 wt% (20 to 26 at. %) Nb, irradiated to about 0.8% burnup of all atoms in pressurized water, exhibited large volume increases when the centre temperature exceeded 360°C (Thomas et al., 1958). Uranium-rich uranium-zirconium alloys behave similarly and do not resist swelling any better than unalloyed uranium (Bauer et al., 1958; Kittel et al., 1964; Pugh and Butcher, 1964). In both sets of alloys the matrix was probably alpha uranium, so the poor performance is not surprising. Specimens of U–10wt% Nb irradiated to nearly 1% burnup of all atoms at temperatures below 200°C and subsequently annealed showed no significant swelling until the eutectoid temperature (635°C) was exceeded (Thomas et al., 1958). Again, the result is what would be expected of a fuel that is basically alpha uranium.

The ternary alloy U-5wt% Zr–1.5wt% Nb is said to have improved resistance to swelling (Reinke and Carlander, 1960; Kittel et al., 1964). The evidence in support of the statement is that specimens irradiated to 0.35% burnup of all atoms at centre temperatures between 300 and 400°C suffered negligible swelling in post-irradiation annealing up to 500°C. Since the exaggerated swelling around 450°C, due to the growth of alpha-uranium grains, occurs only during irradiation, this alloy has not been distinguished from unalloyed uranium by these results. However, U-10.6wt% Nb–4wt% Zr, U-10.6wt% Nb–5wt% Zr, and U-10.6wt%Nb–6wt%Zr alloys showed no large swelling for centre temperatures *during irradiation* up to 450°C in the same series of pressurized-water tests as for the uranium-niobium binaries (Thomas et al., 1958). It seems possible that with such relatively large alloying additions the continuous matrix is gamma phase.

Technologically, the most important discovery was that the small alloying additions of iron and aluminum used to control the grain structure in "adjusted" uranium also served to reduce swelling (Bellamy, 1962). Specimens containing 200 to 500 ppm Fe and 500 to 1200 ppm Al were vacuum melted, then beta quenched and alpha annealed. These showed volume increases of less than 10% when irradiated at 575°C to a burnup of 0.7%. Although it is now known that this temperature is well beyond that giving maximum swelling, control specimens without the additives and others with 0.5 at. % Ti or 0.5 at. % Nb irradiated under the same conditions exhibited severe and even catastrophic swelling. Metallography confirmed that the specimens with small volume changes contained a fine dispersion of porosity, while the others had larger pores more widely spaced: In both, there were grain-boundary cracks, but in the latter they were expanded more, perhaps due to gas pressure.

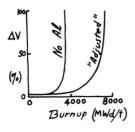

Later results have shown that at high fissioning rates (> 15MW/tonne) "adjusted" uranium suffers a maximum swelling at temperatures around 450°C, just as unalloyed uranium does (Barnes et al., 1964). The important difference is that in the latter breakaway (> 10% volume increase, say) occurs at 3000 MWd/tonne, while in the former it is around 6000 MWd/tonne. The swelling of "adjusted" uranium was found to be unaffected by severe thermal cycling from an upper temperature of 575°C, for irradiations up to 6000 MWd/tonne: Under similar conditions, thermal cycling caused increased volume changes in uranium with no aluminum additions.

In the amounts used in "adjusted" uranium, the alloying additions are largely soluble in the beta phase. On quenching they are retained in a supersaturated solution, but very finely dispersed particles of UAl_2 and U_6Fe are precipitated during alpha annealing. The fact that in Bellamy's experiments only the quenched and annealed specimens exhibited resistance to swelling suggests that the benefit is due more to the precipitates than to the alloying atoms remaining in solution. However, the small amounts of iron and aluminum left in solution may act like molybdenum in decreasing the mobility of xenon. Often the particles appear to be too coarsely distributed to affect the nucleation of pores, so they may hinder pore growth by anchoring them. Since observations by transmission electron microscopy reveal few of the pores associated with precipitated particles, a less direct effect is indicated (Barnes et al., 1964): The particles may serve to anchor dislocation networks and grain boundaries on which the pores are trapped.

One way or another, the particles may well immobilize the pores. However, many of the experimental results showing the advantage of using "adjusted" uranium have been obtained in the temperature range where grain-boundary tearing due to irradiation growth is the predominant mechanism causing damage. In this context, the particles may be helping to prevent grain-boundary sliding, while the finer grain size in "adjusted" uranium may also be helpful. Extraction replicas from polished sections of unirradiated "adjusted" uranium showed a high density of the precipitate particles in the grain boundaries (Makin et al., 1962). The appearance (Fig. 2–14) suggests that the second phase initially precipitated as rods, then spheroidized into particles during the alpha anneal.

The success of "adjusted" uranium in reducing swelling encouraged Kramer et al. (1965) to test the effect of finely distributed precipitates in uranium-molybdenum alloys. They used U-4wt%Mo-0.1wt%Si and U-10.5wt% Mo-0.04wt%Sn alloys, heat treated to produce suitable precipitates, together with uranium-molybdenum binaries to act as controls. These specimens were exposed to approximately 0.5% burnup at centre temperatures of 300 or 400°C. Because of the low irradiation temperatures none of the specimens suffered much swelling, but subsequent annealing differentiated between them. A U-4wt% Mo-0.1wt%Si specimen that originally had 4×10^{12} particles/cm³ decreased in density by only 7.5% after progressive anneals culminating with 24 h at 850°C, while its binary control swelled more after 20 h at 650°C: The U-10.5wt%Mo-0.04wt%Sn specimen with 2×10^{13} particles/cm³ initially had decreased in density by less than 2% after 24 h at 850°C, while its binary control had more than twice as much swelling after the same annealing esquence.

Although the previous results were obtained from post-irradiation annealing, in another series of experiments

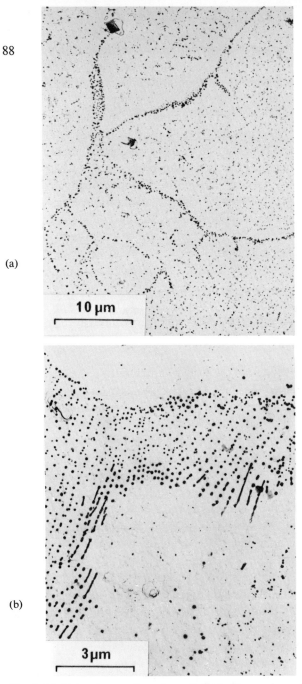

(a)

(b)

Fig. 2–14. Extraction replicas from "adjusted" uranium showing
(a) precipitates predominantly at grain boundaries,
(b) fine structure of precipitates at a grain boundary.
      (After Makin et al. (1962), courtesy Inst. of Metals.)

similar ternary alloys demonstrated a high resistance to swelling even when irradiated with time-average surface temperatures in the range 400 to 500°C (Ballif, 1962). His U-3.5wt% Mo alloys, containing up to 2wt% of either aluminum or silicon to produce finely dispersed precipitates, decreased in density by only about 5% at the lower temperatures and about 12% at the higher temperatures by 15 000 MWd/tonne U. Such behaviour is greatly superior to any seen in unalloyed or "adjusted" uranium, but it is not yet clear how much of the benefit is due to the minor ternary additions.

Thus, for thermal reactors requiring fuel centre temperatures up to 600°C and burnup around 5000 MWd/tonne, "adjusted" uranium has as good resistance to swelling as any of the alloys examined, and it has, by far, the best combination of low parasitic neutron absorption and high uranium density. Barnes et al. (1964) showed that specimens of "adjusted" uranium and U–2.5wt%Mo (6 at.%Mo) irradiated at 500 to 650°C exhibited similar swelling behaviour up to 6500 MWd/tonne. This burnup, however, probably represents the present limit for "adjusted" uranium with centre temperatures over 400°C. U-3.5wt%Mo alloys, possibly containing minor ternary additions, permit the limit to be extended over 10 000 MWd/tonne, at least for temperatures up to 500°C. Operation at temperatures over 660°C is usually avoided because of the large stresses generated by volume changes when alpha uranium transforms to beta uranium. "Adjusted" uranium probably offers no advantages at higher temperatures, since the precipitates dissolve or coarsen in the beta phase.

Uranium-molybdenum alloys, at temperatures up to 650°C, have been used for initial charges in both the Dounreay and Enrico Fermi fast reactors; 9 and 10 wt%, respectively. However, the burnup currently available without excessive swelling (about 2% burnup) is inadequate for most fast-reactor fuel cycles.

Finally, the only plutonium-rich alloy for which there is irradiation experience is Pu-1.25wt%Al (10at.%Al).

For a specific application, the Mark-IV loading of the EBR-I fast reactor, a highly concentrated plutonium fuel was required. Because unalloyed plutonium has four-phase transformations involving large volume changes between room temperature and its melting point at 640°C and because the three phases present below 320°C are brittle, aluminum was added to stabilize the face-centred delta phase normally present only at elevated temperatures. Specimens of the alloy showed very little dimensional change after exposure to 0.09% burnup of all atoms at a fuel centre temperature near 385°C, but these conditions do not constitute a severe test (Storhok, 1963).

2.8 Physical and Mechanical Properties

Irradiation damage to the physical and mechanical properties of uranium is qualitatively similar to that in other metals. However, since the kinetic energy of the fission fragments is about one hundred times that of a fast neutron, the damage for a given flux is much greater in uranium. Thus, serious damage can occur within a few hours in a typical reactor. Also, the thermal neutron flux, which has no effect on most metals, is usually most important in causing fission damage.

Post-irradiation measurements of the thermal conductivity at 60°C have demonstrated significant decreases due to irradiation (Garlick and Shaw, 1965). Bars irradiated with an average temperature of 300 to 450°C had suffered a decrease of about 7% before 100 MWd/tonne (0.01% burnup), and a further 2 or 3% by 3000 MWd/tonne (0.3%): For irradiation temperatures from 200 to 300°C, the decrease was about 15% in the range 1000 to 2000 MWd/tonne. These results agree with earlier ones reported by Billington (1955) but are hard to reconcile with an observation by Bates (1958) that uranium irradiated to 300 MWd/tonne at a maximum temperature of 220°C showed no decrease in conductivity greater than his experimental error of 5%.

Measurements made during irradiation also showed a small decrease in the thermal conductivity of uranium (Billington, 1955). In U-1.6wt% Zr (4at. %Zr) decreases up to 10% were observed in post-irradiation measurements after 7000 MWd/tonne at maximum temperatures of 315°C (Deem et al., 1955). Where a large amount of swelling is encountered, the thermal conductivity of the fuel both during and after irradiation should be appreciably decreased by the porosity developed.

Early post-irradiation measurements showed that the room-temperature electrical resistivity of uranium irradiated to 6×10^{18} fissions/cm^3 (0.015% burnup) at 80°C was increased by a few percent (Konobeevsky et al., 1955). Holding at temperatures around 200°C caused annealing of the damage, but a variable activation energy for the recovery suggested that the process was complex. Quéré and his colleagues have since studied the increases in resistivity for both annealed and worked uranium after irradiation at ambient temperature, −196, −253, or −269°C (Quéré and Nakache, 1959; Quéré, 1963 a&b; Adda et al., 1964; and Burger et al., 1965). After allowing for deformation due to irradiation growth, they found that the resistivity changes tended to saturate at low exposures, of the order of 10^{-4}% burnup. Much of the damage could be explained by supposing that each fission spike produced a large number of point defects, but exact agreement with the experimental results required consideration of defect clustering and the formation of dislocation loops. Damage introduced at −269°C began to anneal below −253°C, i.e., the same range as for unirradiated quenched specimens.

Testing uranium specimens after irradiation has shown that the hardness and yield strength are significantly increased by exposure to only 100 MWd/tonne at temperatures of 150°C and below: The ultimate tensile strength and especially the ductility are concurrently decreased (Paine and Kittel, 1955; Konobeevsky et al., 1955; Madsen, 1956; Bement, 1959; Lee et al., 1955; Bush, 1957). Annealing at temperatures from 400 to 600°C largely restored the strength but left the ductility seriously impaired: This suggests that some of the original damage was due to cracks. Specimens irradiated from 200 to 850 MWd/tonne at centre temperatures around 400°C suffered similar effects (Shaw, 1960).

Measurements made by Zaimovsky et al. (1958) during irradiation in a flux of approximately 10^{13} n/(cm^2 sec) and at temperatures from 150 to 220°C are in marked contrast (Fig. 2–15). Although the yield strength is again increased,

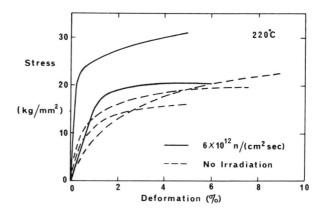

Fig. 2–15. Effect of irradiation in a flux of 6×10^{12} n/(cm^2 sec) on stress/strain curves for uranium. (After Zaimovsky et al. (1958), courtesy United Nations.)

the ductility is only marginally decreased, and the ultimate strength is now increased. These results, like the in-reactor creep results (Sec. 2.4), demonstrate that uranium under irradiation can be ductile. Sykes and Greenwood (1965) have since shown that even in post-irradiation tests the ductility of uranium with 0.05% burnup was comparable to that of unirradiated material, if the stress was reduced below a critical value (3.5 kg/mm^2 at 525°C). They supposed that the low stresses were insufficient to cause coalescence of pores on grain boundaries: Annealing between irradiation and testing *decreased* the ductility at low stress, presumably by permitting pore growth.

2.9 Metallurgical and Crystal Structure

Metallography, in a variety of forms, has been exploited extensively to study the effects of irradiation in uranium. Some of the observations have already been described in connection with specific phenomena; for instance, the dislocation loops that are the basis for an explanation of irradiation growth and the porosity associated with swelling.

In uranium irradiated below 200°C, increased twinning was seen after only 0.005% burnup (Bierlein and Mastel, 1958). As the exposure was increased, the existing twins widened and fresh ones appeared, some intersecting existing ones. Grain rotation also occurred and minute transgranular cracks could be observed. By 0.1% burnup the material was heavily twinned, the twins were curved, and grain boundaries were no longer sharply delineated (Fig. 2–16). Similar twinning was seen in uranium irradiated to 0.2 to 0.6% at temperatures between 200 and 300°C (Eldred et al., 1958; Bloch et al., 1958). Examination of a cross section of a bar of U-0.5wt%Mo (1.2 at.% Mo) that had been irradiated to 1.6% burnup showed that the metallographic structure was much more disrupted in the outer region at 430°C than in the centre at 580°C (Englander et al., 1964).

These effects presumably result from deformations in grains whose irradiation growth is frustrated by adjacent, misoriented grains (Sec. 2.3). By 580°C, growth is very slight, and the small amount of deformation observed may have occurred during thermal cycles. In U-1.5wt%Mo (3.7at.%Mo) specimens, where the gamma phase was present as lamellae in the alpha grains, irradiated to 0.3% burnup, severe rumpling of the lamellae for irradiation temperatures below 600°C indicated the flow pattern of the grains: At 600°C, the lamellae were not distorted but became spheroidized (Englander et al., 1964).

Bloch et al. (1958) found that post-irradiation annealing in the alpha phase caused coarsening of the twinned structure, but that recrystallization did not occur until the

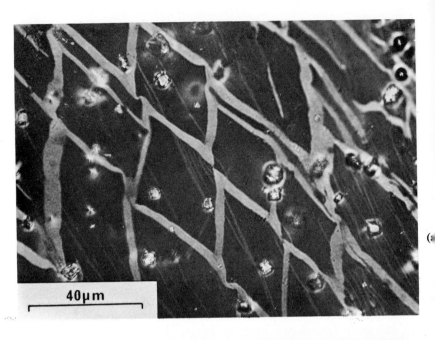

(a

Fig. 2–16. Microstructures of irradiated uranium.
(a) Twinning in unalloyed uranium 600 MWd/tonne at 240°C.
(After Eldred et al. (1958), courtesy United Nations.)
(b, c, d) U–0.5wt%Mo irradiated to 1500 MWd/tonne.

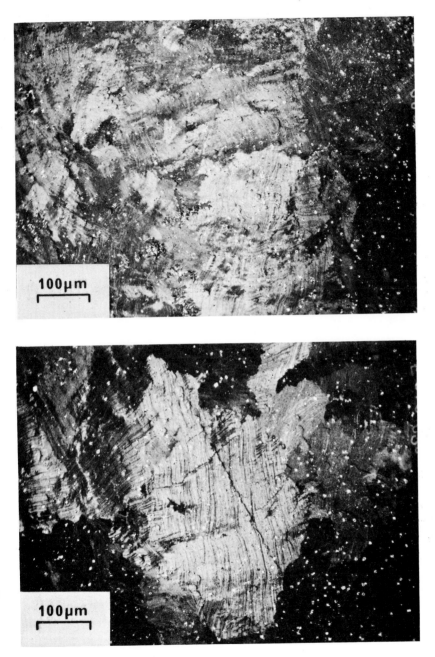

(b) outer edge at 430°C,
(c) middle annulus at 500°C,
(d) centre at 580°C.
  (After Englander et al. (1964), courtesy United Nations.)

specimen was heated into the beta phase. Irradiation suppressed recrystallization that would otherwise have been caused by thermal cycling (Colabianchi et al., 1964). The observations of recrystallization in the alpha phase during irradiation by Loomis and Pracht (Sec. 2.5) are probably a result of the high purity uranium they used. Englander et al. reported that they had never observed any evidence of grain growth having taken place in uranium during irradiation even at 600°C. Where the centre temperature exceeded the alpha-beta transformation temperature, however, the result was a coarse columnar structure with many cracks.

Using transmission electron microscopy, Makin et al. (1962) saw dots about 50- to 100-Å diam in uranium irradiated at 80°C to an exposure of 2×10^{16} fissions/cm^3 ($5 \times 10^{-5}\%$ burnup). The dots lay on lines and preferentially at grain boundaries. By 2.4×10^{17} fission/cm^3 the defects were more uniformly distributed and could be identified as dislocation loops, while subsequent annealing at 300°C increased their size up to 0.1-μm diam (Fig. 2–17). Both transmission and replication techniques revealed the the presence of fine pores. In uranium that had been irradiated at 530°C, polyhedral pores over 1 μm in diam were observed at grain boundaries. The surfaces of several exhibited a terraced structure indicative of appreciable surface diffusion, and many had linked together. The density of pores varied widely from one boundary to another in an apparently random manner. The consequent weakening of the boundaries, together with strengthening of the grain interiors by dislocation pinning, probably accounts for the predominantly intergranular fracture of irradiated uranium.

Irradiation at temperatures under 100°C produces line broadening in the x-ray diffraction pattern of annealed uranium but a reduction of line breadth in cold-worked metal Paine and Kittel, 1955; Konobeevsky et al., 1955; Bierlein and Mastel, 1958; Tardivon, 1959). Presumably the point defects produced by fission not only distort the lattice but also assist in dislocation climb. The line

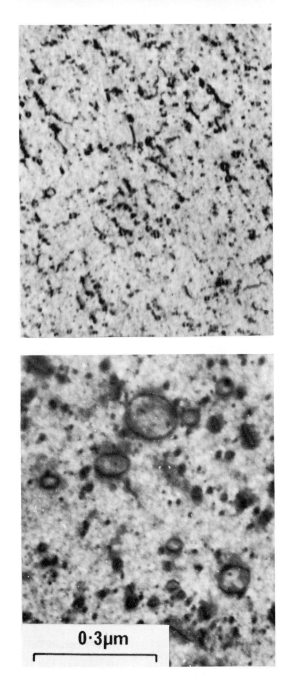

Fig. 2–17. Dislocation loops in uranium irradiated to 1.6×10^{17} fissions/cm³ at 80°C.

(a) As-irradiated.

(b) Annealed one hour at 350°C.

(After Makin et al. (1962), courtesy Inst. of Metals.)

broadening saturates around $10^{-3}\%$ burnup. Irradiation causes very little change in the lattice parameter of alpha uranium: Indeed, those reporting changes in line breadth usually do not bother to comment on the lattice parameter.

2.10 Corrosion

Unalloyed uranium has such poor resistance to aqueous corrosion that little attention has been given to the effect of irradiation on the corrosion rates. Although there is experience of uranium fuel elements rupturing in water-cooled reactors, the ill-defined geometry and unknown effects of the sheath prevent interpretation in terms of corrosion rates per unit surface area. Even outside the reactor, the rate at 300°C is about 10^5 mg/(cm^2 day), or, expressed as a penetration, approximately 0.001 in./min. Kittel et al. (1957) showed, in one specimen, that the post-irradiation corrosion rate at 100°C is little affected by a burnup of 0.1% of the uranium.

Thus, most of the information on in-reactor or post-irradiation corrosion relates to alloys that have been previously selected for their resistance to corrosion. Alloying elements investigated include molybdenum, niobium, and zirconium (uranium-silicon alloys are considered in Sec. 4.7). They have normally been prepared so they retain the gamma phase, or as supersaturated alpha phase, since these conditions provide the best resistance. In normal corrosion of uranium alloys a protective oxide grows on the metal surface, but a nonprotective one further out keeps sloughing off. However, a much more serious form of corrosion, apparently due to hydride formation in the metal, can occur abruptly; the nature of the attack depending on impurities, working, and texture. Sheaths can therefore reduce the corrosion not only by decreasing the area attacked but also, when made of a metal such as zirconium, by absorbing hydrogen (Burkart, 1956). Conversely, the rates can be greatly accelerated in steam due to the heat of the corrosion reaction not being rapidly dissipated and thus causing temperatures as high as 700°C locally (Troutner, 1960).

Jones (1957) irradiated specimens containing 9 to 13.5 wt% (20 to 28 at.%) Mo or 10 wt% (22 at.%) Nb, then he conducted corrosion tests in static autoclaves at 343°C. For exposures up to 0.2% burnup of all atoms, the molybdenum alloys exhibited no significant changes in the corrosion rates, but the time to failure was reduced by an order of magnitude to a few days. The corrosion rates of the irradiated niobium alloys were markedly accelerated up to the maximum exposure tested, 0.09% burnup of all atoms. The authors suggested that the in-reactor corrosion rates could be lower for those alloys in which the gamma phase is stabilized by irradiation (Sec. 2.6).

Post-irradiation corrosion tests have also been performed on specimens containing lower amounts of alloying addition. Kittel et al. (1957) reported that irradiation destroyed the corrosion resistance of U–3wt%Nb (7 at.% Nb) and U–5wt%Zr–1.5wt% Nb (11 at.% Zr–4 at.% Nb) by an exposure of 0.1% burnup of all atoms. Greenberg and Draley (1958) confirmed these conclusions and extended them to include U–3 wt% Nb–0.5 wt% Sn (7 at.% Nb–1 at.% Sn). Later, Greenberg (1959) published results that apparently contradicted the earlier ones. A plate was removed from a fuel element irradiated in the Experimental Boiling Water Reactor and samples with different burnup taken for corrosion testing at 260 to 270°C. The U–5 wt% Zr–1.5 wt% Nb alloy showed a corrosion rate that initially decreased with increasing exposure, reaching a minimum of about one fifth the pre-irradiation value near 0.05% burnup of all atoms, then increased again by 0.14%. However, these samples appear to have been tested still partly sheathed; a circumstance that may explain the surprisingly good results. Kittel and Smith (1960) found that the corrosion resistance at temperatures around 300°C of U–2 wt% Zr (5 at.% Zr), coextruded with Zircaloy–2 and then heat treated to promote interdiffusion, was qualitatively unchanged by exposures up to 0.1% burnup.

Thus, alloying and sheathing can improve the corrosion resistance of uranium, but there remains the suspicion that irradiation probably causes some deterioration.

Although little has been published on the subject, irradiation apparently does not seriously affect the corrosion of uranium in carbon dioxide at temperatures up to 500°C (Pugh and Butcher, 1964). At 800°C, however, Parker et al. (1960) found that the corrosion in air, carbon dioxide, and steam was greater for irradiated than unirradiated material, possibly because the release of fission-product gases disrupted the protective oxide film. In all oxidizing media very rapid rates are to be expected if uranium that has experienced much swelling becomes exposed, simply because of the greater accessible surface area. Indeed, there is concern that in extreme cases the fuel could become pyrophoric.

Oxide Fuels

3.1 Uranium-Oxygen System

Around 1954, interest in uranium dioxide began to develop rapidly. Initially, its excellent resistance to aqueous corrosion was the stimulus, but its dimensional stability under irradiation was also appreciated as experience of swelling in metal fuels became widespread. Since then, much work has been done on the fabrication, properties, and testing of uranium dioxide — in 1961, a comprehensive review volume edited by Belle cited nearly 1000 references — and it has come into general use as a fuel for power reactors. Greater familiarity with the material has shown how naive it was to regard uranium dioxide as a simple, inert compound, and there are still many unresolved problems concerning its behaviour.

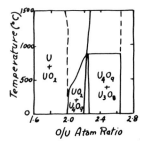

The equilibrium diagram for the relevant portion of the U-O system is sketched alongside. At the stoichio-metric composition of UO_2 the crystal structure is the well-known fluorite (CaF_2) one, in which the uranium atoms lie on a face-centred cubic lattice with the oxygen atoms in the spaces between. At elevated temperatures, excess oxygen can be dissolved in the UO_2 structure by oxygen atoms going into interstitial sites. Neutron diffraction showed that not only the excess oxygen atoms but some of the others too occupy interstitial sites (Willis, 1963a). Subsequent work by Willis (1963b) indicated that in stoichiometric UO_2 either an appreciable fraction of the oxygen atoms were interstitials at elevated temperatures or they vibrate anharmonically. For a composition U_4O_9,

O = o   ● = U

the stable structure is one in which the excess oxygen atoms adopt an ordered arrangement with the unit cell dimensions approximately four times those for UO_2. Although the distinction between the UO_2 and U_4O_9 structures seems a small one, the resulting difference in properties can be marked. For instance, a precipitate of U_4O_9 in UO_2 can be clearly observed metallographically.

Most equilibrium diagrams published show no solubility for uranium in the UO_2 structure, but recent work indicates that, at least at temperatures around 2000°C, appreciable amounts of uranium can dissolve without a change of structure. On cooling, precipitation of uranium indicates a reduction in solubility with decreasing temperature. The partial pressure of oxygen in equilibrium with stoichiometric UO_2 is very low, even at high temperatures. Thus, hypostoichiometric compositions can be obtained (and maintained) only under the most stringently reducing conditions.

The stability of the UO_2 structure over a finite range of compositions means that stoichiometric UO_2 exists only as an ideal concept. In practice, any specimen is to some degree hypo- or hyper-stoichiometric. The nature of the bonding in UO_2 is similarly ill-defined. An ionic structure is often assumed, but there has been no convincing proof of this. If the bonds are predominantly (but not fully) ionic, the degree of their ionic nature may vary with temperature.

Because of its high melting point (2800°C) and brittle behaviour at temperatures below 1000°C, UO_2 is normally fabricated by the methods of powder metallurgy. A powder of large surface area is cold compacted into a pellet, then sintered at temperatures around 1600°C to cause densification. The conditions actually used vary widely, but the product is a polycrystalline material containing a few percent porosity dispersed through the grains and at grain boundaries. Arc melting is also possible, and this route yields a product of very large grain size of the order of one centimetre. The bulk density normally exceeds 99% of that theoretically attainable: Macropores, micropores, and porosity that is resolved only by electron microscopy account for the difference. The arc-melted material is prone to contamination, notably by second-phase nitride particles, but pure UO_2 can be obtained by this process. Other methods of preparation exist but are not widely used for making reactor fuel.

3.2 Structural Changes

The structural changes that occur in sintered UO_2 during irradiation (Fig. 3–1) are striking and can be most interesting. However, they are largely thermal effects that can be simulated in the laboratory. One reason that the structures appear unusual is that the temperatures (over 2800°C for central melting) and especially the thermal gradients (as high as 10^4 deg C/cm) are considerably higher than those encountered in most other fields.

Rods that have been irradiated at relatively low power per unit length (Fig. 3–1a) show only cracks, predominantly radial in character. These are simply attributable to thermal stresses during operation. Classical analysis predicts cracking for

$$\Delta\theta = 2P(1 - \mu)/E\alpha, \qquad (3-1)$$

where $\Delta\theta$ is the temperature difference between the surface and centre of the fuel, P the rupture modulus, μ Poisson's ratio, E Young's modulus, and α the coefficient of linear thermal expansion of the fuel. If values of these properties measured in the laboratory are inserted in Eq. (3–1), the $\Delta\theta$ for which cracks are to be expected is around 100 deg C near room temperature and increases slightly with increasing temperature. In practice, laboratory experiments have confirmed the predictions reasonably well, but apparently uncracked pellets have been observed after irradiations in which the temperature difference was at least twice as much (Robertson et al., 1962). Thus, there is the suspicion of an irradiation effect and, of the relevant properties, the modulus of rupture

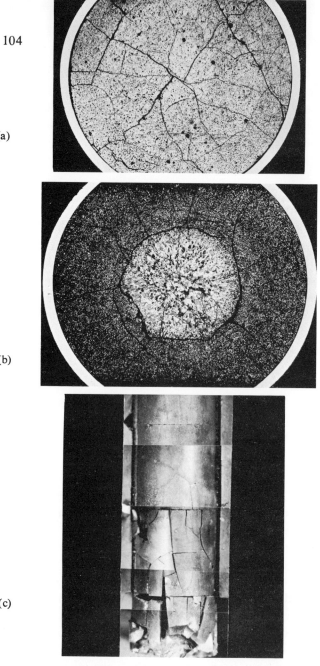

Fig. 3–1. Structural changes on irradiating sintered UO_2.
(a) Polished cross section: Thermal cracking.
(b) Polished cross section: Central grain growth.
(c) Sheath stripped to reveal cracking pattern on pellet surfaces.

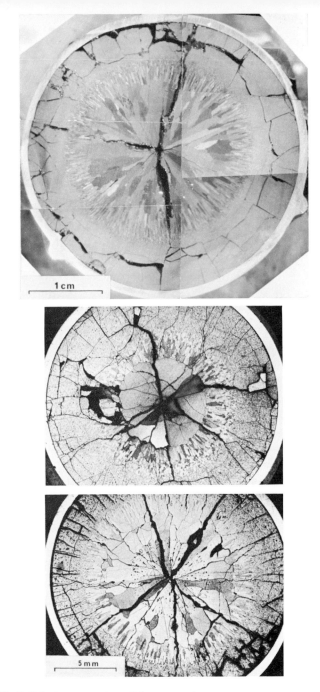

(d) Polished cross section: Columnar grains.
(e) Polished cross section: Very large grains at the centre.
(f) Polished cross section: Central porous region.
(Courtesy A. S. Bain, Chalk River Nuclear Laboratories.)

appears the one most likely to be affected to a sufficient extent.

Similar radial cracks are still seen after an irradiation at higher power per unit length (Fig. 3–1b), but they are now in the outer, cooler annulus of the UO_2. Generally, a circumferential crack contains a central region of enlarged grains. On the outside, curved surface of the pellet the cracks form a predominantly rectangular pattern (Fig. 3–1c).

By artificially dividing pellets to simulate cracking prior to irradiation, Bain (1963) showed that cracks in the inner region healed during irradiation, whereas those in the outer annulus remained unaltered (Fig. 3–2a). Thus, it appears that as the power increases during the initial startup the pellet is cracked into successively smaller fragments by the thermal stresses. During operation, the fragments in a central region above a certain temperature sinter together, possibly under pressure. On a subsequent shutdown, decreasing the thermal gradients in the central region again causes thermal stresses, but this time, in the opposite direction. The central cracks are therefore deduced to form on cooling at each shutdown, in contrast to those in the outer annulus that form once and for all in the initial heating.

Bain observed similar behaviour of deliberately produced cavities in the UO_2 (Fig. 3–2b). Those in the outer annulus remained intact, while those in the hotter central region closed up during irradiation. Inert tungsten wires that had been placed in the cavities did not move from their original positions during the irradiation, but in the central region the wires were no longer loose in the cavities.

Where grain growth is first observed in the UO_2 (Fig. 3–1b) the grains are equiaxed, and the effect can be readily explained by the originally fine-grain material having been held at an elevated temperature for a prolonged period. In laboratory investigations, MacEwan (1962) determined how the final grain size of UO_2 sinters depends on the initial value, the temperature, and the time at temperature. With Robertson and others (1962), MacEwan showed how different batches of UO_2 prepared

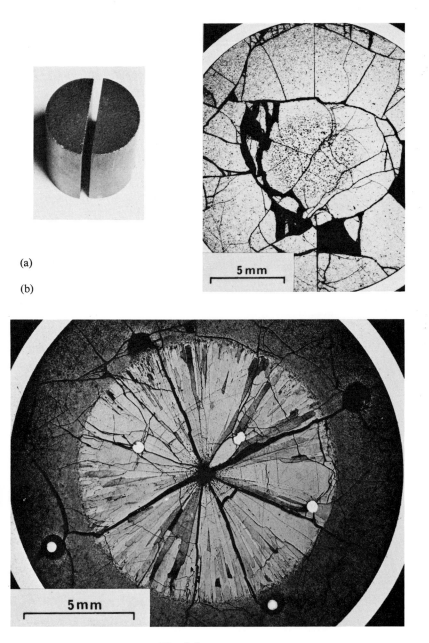

(a)

(b)

5 mm

5 mm

Fig. 3–2.
(a) Composite pellet of UO_2 before and after irradiation, showing
    the healing of a crack in the central region.
(b) Polished cross section of irradiated UO_2 that originally contained
    drilled holes. The white spots are tungsten wires: Some holes
    are empty because wires fell out during examination.
(After Bain (1963).)

in nominally the same manner exhibited appreciable variation in grain-growth characteristics; also, how accelerated grain growth occurred in regions of large stress gradients. Thus, even without irradiation, several factors can affect the grain growth.

By annealing a previously irradiated sample of UO_2 alongside an unirradiated control broken from the same sinter, MacEwan and Hayashi (1965) demonstrated that prior irradiation impedes grain growth. Optical microscopy showed the pores to be mainly located on the grain boundaries of the irradiated sample, while they occurred mostly within the grains of the as-sintered material. The decrease in grain growth is therefore probably due to the pores hindering the movement of grain boundaries. In a subsequent irradiation, the centre temperature of a UO_2 specimen, measured continuously by thermocouple, increased from about 1400 to 1700°C over a period of a month: An unirradiated specimen from the same batch was subjected to exactly the same thermal programme, but in the laboratory (MacEwan and Hayashi, 1965). Examination showed the grain size of the unirradiated specimen to be 37µm while that at the centre of the irradiated one was only 7.5µm. Studying thin films of UO_2 by transmission electron microscopy, Golyanov and Pravdyuk (1964) found the temperature of appreciable grain growth increased from 1600°C for unirradiated material to 1800°C for a specimen irradiated to 10^{18} fissions/cm³ (4×10^{-3} % uranium burnup)* near 100°C. In all these comparisons the UO_2 was first irradiated at a temperature below that for appreciable grain growth. Whether irradiation at a temperature where grain growth is normally rapid has any effect has yet to be established.

Columnar grains are frequently seen inside the ring of equiaxed grains, if the power per unit length is sufficient (Fig. 3–1d). The elongation of these grains in a radial direction suggests that not only the temperature but also the thermal gradient is involved in their formation. Metallo-

---

* In 95% dense UO_2, 1% uranium burnup $\equiv 2.3 \times 10^{20}$ fissions/cm³.

graphic examination of UO_2 sinters from laboratory experiments simulating thermal conditions in the reactor led MacEwan and Lawson (1962) to propose a mechanism: There is a lenticular pore that occupies the full cross-sectional area of the grain at the hot end of many of the columnar grains. The pore is believed to migrate up the thermal gradient by sublimation of the oxide, digesting the polycrystalline matrix ahead of it and continuing the growth of a single crystal on the cooler side.

As the power output per unit length is further increased there is often a region of large grains inside the annulus of columnar grains (Fig. 3–1d). Sometimes the large grains appear without any surrounding columnar grains (Fig. 3–1e). Many of the grain boundaries are oriented radially but the grains are not especially elongated so that although they have been called "columnar grains", the term is strictly a misnomer. It appears that a few grains close to the centre and under near-isothermal conditions grow at the expense of their smaller neighbours until they meet each other. Thereafter, they grow radially outwards, but only as long as their outer limit is at a temperature of appreciable grain growth.

The amount of porosity in the as-sintered material is probably important in determining whether columnar or broad grains form in the hot central region. In one irradiated element that he had sectioned longitudinally, Bain (1963) observed narrow columnar grains in those pellets that had initially contained three percent porosity, and broad grains in those with two percent. Similarly, Notley and MacEwan (1965) noted a tendency for narrow columnar grains to be replaced by broad grains as the initial porosity decreased from five to two percent. However, the final grain structure depends on other factors as yet unidentified; perhaps the morphology of the original pores, the thermal gradient at the position of grain growth, and the thermal history among others. Since narrow columnar grains are often observed in irradiated sections where no lenticular pores can be seen, there is probably at least one other mechanism besides pore migration for the formation of the columnar structure. Any deduction of fuel temperatures from the structural changes is dangerous, since many factors have been shown to affect the different forms of grain growth.

Where the power per unit length has been very high, the fuel exhibits a new central region usually well defined on both fractured and polished sections (Fig. 3–1f). The large grains are continuous across the boundary of this region, but, within, there is a substructure. In relatively pure UO_2, cross sections exhibit a cellular network, while a dendritic structure has been seen in material containing impurities (Fig. 3–3). In the same region there are large pores; some nearly spherical and others more tubular in a radial direction. Since these features are similar to those observed in metals frozen from the melt (Winegard, 1961), it is inferred that the central region has been above the melting point of approximately 2800°C. This interpretation has been supported by experiments with inert markers (Sec. 3.3), measurements of fuel expansion (Sec. 3.5), and autoradiography (Sec. 3.8).

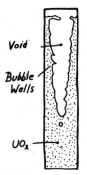

Some UO_2 specimens were exposed to an even higher power per unit length in deliberately provoked reactor power transients (Field et al., 1962). A longitudinal cross section showed a series of large bubbles comparable with the diameter of the fuel and apparently frozen in while rising to the surface. The operating conditions were not well known, but the authors estimated a centre temperature of 4000°C and suggested that the UO_2 had boiled. At much lower temperatures (below 2800°C), substantial material transfer can occur by sublimation of the UO_2. Specimens prepared with axial voids running the full length of the UO_2 have had, after irradiation, the void sealed at both ends with a UO_2 deposit (Bain, 1965). The shape of axial voids that form in solid pellets during irradiation at temperatures below the melting point is probably dictated by volatilization of UO_2 from the hotter areas and deposition on the cooler ones. Figure 3–4 illustrates single crystals, presumably grown from the vapour, that were found inside such a void.

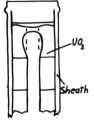

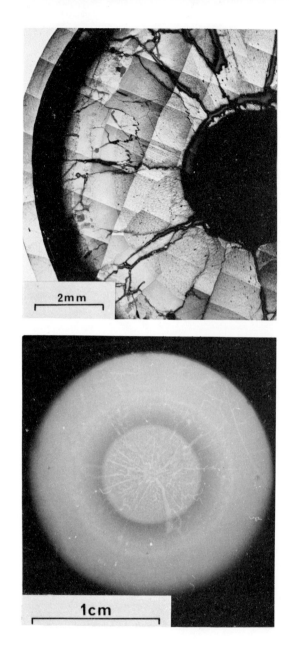

(a)

2mm

(b)

1cm

Fig. 3–3. Substructures seen in regions believed to have been molten during irradiation.

(a) Polished cross section of pure UO$_2$. (Courtesy G. R. Horn, Battelle-Northwest Laboratories.)

(b) Autoradiograph of cross section similar to (a) — white indicates activity. (Courtesy A. S. Bain, R. D. MacDonald, A. D. Murray, Chalk River Nuclear Laboratories.)

112

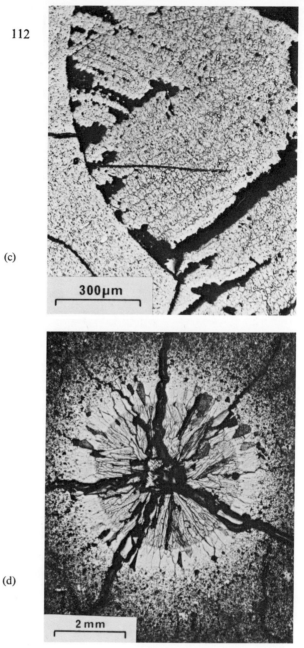

(c)

300µm

(d)

(c) Polished cross section of impure UO₂. (Courtesy R. N. Duncan, Westinghouse Atomic Power Department.)

(d) Polished cross section of ThO₂–19wt%UO₂ (Courtesy G. W. Parry, Chalk River Nuclear Laboratories.)

Fig. 3–4. Crystals found deposited in the central void of an irradiated
UO₂ fuel element. (After A. S. Bain et al. (1964a).)

Structural changes on a finer scale are revealed by electron
microscopy. Replicas of both fractured and polished
surfaces of UO_2 sinters showed pores from 100- to 1000-Å
diam on grain boundaries after an exposure of 10^{18} fis-
sions/cm³ at a temperature estimated to be below 250°C
(Newkirk et al., 1960). This microporosity was additional
to the original, coarser porosity and was generally absent
at higher exposures up to 10^{19} fissions/cm³. The irradiation
caused the fracture to change from predominantly transgranular to intergranular. After exposures around 2×10^{19}
fissions/cm³ (0.1 % uranium burnup), optical microscopy
showed that at temperatures just below those for discernible

grain growth the original porosity within grains had largely been replaced by pores at the grain boundaries. It is not clear whether the pores moved as such or by the migration of individual vacancies. After still higher exposures, around 10^{21} fissions/cm^3, and at temperatures below those for grain growth, the initial porosity had decreased to a minimum (Padden and Schnizler, 1962). Electron microscopy (Fig. 3–5a) showed enhanced etching of the grain boundaries, possibly because of fission-product segregation (Sec. 3.8). Figure 3–5b shows how, when exposures reached 3×10^{21} fissions/cm^3, the original grains had subdivided into a substructure whose unit size was under 1 μm, and a uniform distribution of fine pores had appeared (Bleiberg et al., 1962). These pores grow, becoming visible by optical microscopy, and are responsible for swelling of the UO_2 (Sec. 3.6).

Returning to lower exposures, individual fission tracks can be seen by transmission electron microscopy. When a thin film of polycrystalline, vacuum-deposited UO_2 was irradiated the tracks were apparent by the furrows they left on the surface (Whapham and Makin, 1962). The tracks were obvious in material of grain size 28 Å, barely visible at 100 Å, and not to be seen at 150 to 500 Å. Whapham and Makin suggested that the effect was due to grain boundaries and free surfaces reflecting energy being dissipated from the tracks. Bierlein and Mastel (1960) found discernible tracks in thin films after an exposure of 3×10^{16} fissions/cm^3, while by 4×10^{18} fissions/cm^3 grain growth had occurred and the UO_2 had agglomerated into isolated filaments. The presence of air during irradiation accelerated the process. In large-grain material tracks can be revealed by diffraction contrast. There was a series of spots along most of the tracks, probably defect clusters, with an average separation around 1000 Å and about 100-Å wide by 220-Å long (Whapham and Makin, 1962; Blank and Amelinckx, 1963).

Dislocation loops around 100-Å diam have also been observed by transmission electron microscopy (Blank and Amelinckx, 1963; Golyanov and Pravdyuk, 1964). Stoichio-

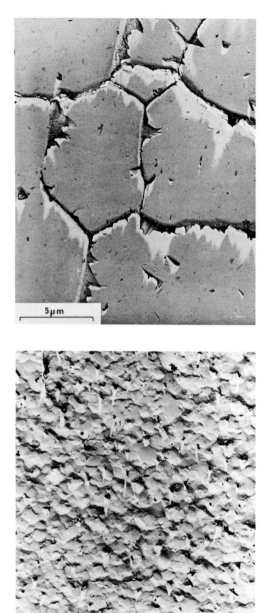

(a)

(b)

Fig. 3–5. Electron micrograph of negative replicas from irradiated sintered UO_2. (After Bleiberg et al. (1962), courtesy International Atomic Energy Agency.)

(a) 1.3×10^{21} fissions/cm^3; etched with 1% H_2SO_4 in H_2O_2.

(b) 3.1×10^{21} fissions/cm^3; etched with a glow discharge.

metric UO_2 specimens irradiated at some temperature below 800°C contained about 3×10^{15} loops/cm^3 after 10^{16} fissions/cm^3 and over 3×10^{17} loops/cm^3 after 10^{19} fissions/cm^3. Many 20-Å-diam pores were situated on the grain boundaries after 10^{18} fissions/cm^3. A specimen that had been irradiated to 10^{18} fissions/cm^3 was subsequently annealed at successively higher temperatures: At 900°C, the loops began to anneal away, and a uniform distribution of very fine spots, about 10-Å diam, appeared; at 1100°C, the loops had almost all gone, the grain-boundary pores had increased in diameter to about 50 Å, and there were about 10^{18} fine pores per cubic centimetre; at 1500°C, the pores in grain boundaries had grown to 300-Å diam, and those in the grains to about 100 Å.

Other investigators have substantially confirmed these observations and extended them to higher exposures (Whapham and Sheldon, 1963; Whapham, 1965). Interstitial loops were seen in sintered UO_2 that had been irradiated to 3×10^{17} fissions/cm^3, and, as the exposure was increased to 2×10^{19} fissions/cm^3, the loops expanded to form networks. When some of the material exposed to the longer exposure was subsequently annealed at 1100°C, pores about 50 Å in diameter appeared between the dislocations: After one hour at 1500°C, the pores had doubled in size and attained a density of 5×10^{15}/cm^3. Pores on dislocations and grain boundaries were larger. Up to 1800°C, there was little change in the structure, but, from 1800 to 2000°C, the grain boundaries were mobile and appeared to sweep the pores so that by 2000°C the boundaries were largely covered by pores about 5000 Å in diameter. In another batch of UO_2, irradiated to a higher exposure (1.5×10^{20} fissions/cm^3) then annealed at 1100°C, dislocation loops were again seen. At 1500°C, the loops coarsened into a network: Pores about 1000 Å in diameter lay on the dislocations, while others about one tenth the size occurred in clusters between the dislocations. Precipitates of the solid fission products, up to 1000-Å diam, were associated with many of the pores.

Whapham re-irradiated one of his specimens that had been exposed to 2×10^{19} fissions/cm^3 then annealed at 1600°C. A further 2×10^{19} fissions/cm^3 caused the 100-Å-diam pores to disappear, leaving only blurred regions of strain. Such direct observations of the behaviour of pores are of great value to the interpretation of gas release (Sec. 3.6) and swelling (Sec. 3.7). However, they also suggest that the irradiation damage in any particular UO_2 specimen may depend markedly on its irradiation history and pre-existing imperfections. An observation by Amelinckx et al. (1965) that the temperature of irradiation has a profound effect is of great significance: An increase from 80 to 300°C greatly increased the density of observable damage in UO_2 single crystals exposed to 10^{16} fissions/cm^3. At 80°C, there were only a few dislocation loops with diameters up to 100 Å, while at 300°C there was a dense pattern of spots with diameters as little as 30 Å.

3.3 Melting

Reliable identification of UO_2 that has been molten during irradiation is important to conclusions regarding the temperature distribution (Sec. 3.4). Confirmation of the interpretation proposed in Sec. 3.2 was obtained from elegantly simple experiments by Horn et al. (1963). In one, they irradiated vertically an element containing UO_2 pellets in which small tungsten marker spheres were distributed. On subsequent examination, the markers originally in the characteristic central region of the fuel were no longer apparent on a cross section near the mid-length. They were, however, found near the bottom of the fuel where they had presumably sunk when the centre was molten (Fig. 3–6). The central region, as defined by the structural appearance described in Sec. 3.2, probably represents the UO_2 that was cooled rapidly from the melt at the final shutdown. Lyons et al. (1964a) suggested that UO_2 solidifying slowly during the irradiation does not

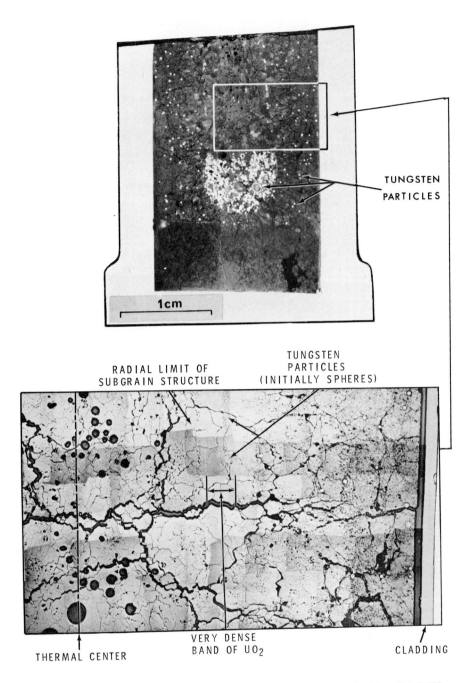

TUNGSTEN
PARTICLES

RADIAL LIMIT OF
SUBGRAIN STRUCTURE

TUNGSTEN
PARTICLES
(INITIALLY SPHERES)

THERMAL CENTER

VERY DENSE
BAND OF UO₂

CLADDING

Fig. 3–6. Longitudinal section from bottom end of irradiated UO₂ containing tungsten particles: Inset at higher magnification. (After Horn et al. (1963).)

produce the characteristic structure of subgrains and pores but, rather, a featureless pore-free annulus. Close inspection of Fig. 3–6 confirms that the limit of melting extends beyond the substructure and, probably, to the outer limit of the pore-free band. Thus, the "high-water mark" is somewhere outside the central region of the structure, by an amount that depends on the ratio of maximum-to-final centre temperature.

Although the melting point of unirradiated UO_2 is well established, the effect of irradiation is still in doubt. From visual observations on fragments of UO_2 on resistance-heated, tungsten-strip filaments (Mendenhall filaments), Christensen (1962) concluded that prior irradiation affected the material's melting point. Samples of unirradiated UO_2 were seen to melt at 2790 ± 20°C; a value in good agreement with the findings of other investigators. Samples of UO_2 taken from a single batch and all irradiated at temperatures under 100°C in the same reactor to exposures of 0.005, 0.014, 0.021, and 0.051% burnup of the uranium exhibited melting points that progressively increased to 2920°C. Six samples from other batches irradiated at various ill-defined temperatures to exposures in the range 0.09 to 0.4% burnup appeared to melt at intermediate temperatures, but these showed no clear dependence of melting point on exposure. By 1% burnup the melting point had been restored to its unirradiated value, while by 8% it had fallen to 2760°C and by 11% to 2660°C.

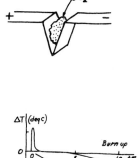

In a second series of experiments, Christensen et al. (1964) determined melting points in a similar apparatus, but with the helium atmosphere static instead of flowing. As the burnup increased from 0.25 to 6% of the uranium, the melting point decreased by 120 deg C, a result in general, but not exact, agreement with the earlier work. None of the specimens in the second series was at a sufficiently low burnup to confirm or deny the melting-point maximum, seen previously.

The decrease at high exposure can be attributed to the impurity effect of the fission products, which by then would constitute 10 to 20 at.% of the uranium. However, the maximum at low exposures is less easily explained. The experiments were well controlled with irradiated and unirradiated samples heated alongside each other for direct comparison. Variations in the irradiation temperature of some samples, by affecting the fractional release of certain fission products, might be thought to contribute to the scatter in melting points, but repeated cycling through the melting point produced no noticeable change in the value for some high melting samples. It is conceivable that even a light irradiation could somehow affect the emissivity of the sample's surface and thus change the apparent temperature of melting by this method. It would, therefore, be reassuring to have the observations confirmed by an independent method. Christensen (1963a), using radiography, measured the volume change on melting unirradiated UO_2 in a resistance-heated tungsten capsule; Hausner (1965) determined the freezing point of unirradiated UO_2 by a thermal-arrest method. Either technique could be adapted to determine the melting point of irradiated material.

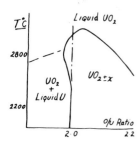

Christensen offered two possible explanations: If it is supposed that the melting point of unirradiated UO_2 passes through a maximum at a slightly hyperstoichiometric composition (say $UO_{2.006}$) and if the fission products require less oxygen than did the uranium from which they formed, then the melting point of irradiated material should pass through a maximum with increasing exposure. He suggested that the difference between the two sets of results was due to the variation in atmosphere affecting slightly the stoichiometry of the uranium oxide. This matter ought to be resolved by comparing samples from the same specimen under both conditions. Alternatively, Christensen argued from a consideration of the energy as a function of entropy that a high density of imperfections in the lattice would enable the material to exist as a super-heated solid at temperatures above its normal melting point. The author justifiably doubted that individual vacancies produced during irradiation would persist to the melting point, but vacancy/gas clusters might provide the requisite defects. If either explanation is valid, the melting point of material being irradiated might well differ from that measured after irradiation.

Those irradiated samples that exhibited high melting points volatilized less rapidly than did their unirradiated controls. However, when the latter melted, giving improved thermal contact with the filament, their temperatures appeared to rise above those of the still solid irradiated samples. Thus, the volatilities at the same temperature of the two materials need not necessarily have differed.

3.4 Heat Transfer

To the extent that phenomena occurring in UO_2 during irradiation result simply from the temperature distribution and do not depend on the irradiation *per se*, the thermal conductivity of the material is very important. The most commonly quoted disadvantage of UO_2 is its low conductivity: The values measured are among the lowest obtained for crystalline solids and are only slightly greater than those for glasses. However, in a refractory material like UO_2, relatively high temperatures are required to produce any given effect, so a low conductivity can be tolerated.

Factors affecting the thermal conductivity of UO_2 have been reviewed (Robertson et al., 1962; Christensen, 1963b). As with other nonmetals, the electronic contribution to the thermal conductivity is normally negligible and lattice conductivity preponderates. Thus, increasing temperature, by decreasing the lattice periodicity, reduces the conductivity. A lower limit of roughly 0.015 W/(cm deg C) can be calculated as the conductivity corresponding to phonon scattering at every lattice spacing. A curve of the conductivity of unirradiated material as a function of temperature, derived from some of the more reliable determinations, is reproduced in Fig. 3–7. A few recent determinations indicate that the conductivity begins to increase again as the temperature exceeds approximately 1500°C, but measurements at such high temperatures are subject to relatively large errors. Impurities also serve to decrease the lattice periodicity and, hence, reduce the conductivity. Excess oxygen in the UO_2 structure, forming a hyperstoichiometric composition, is a special case of an impurity that reduces the conductivity. Although, judged as a solid, the conductivity of UO_2 is low, it is considerably higher than that of a gas. Pores therefore diminish the conductivity of a fully dense sample, with the magnitude of the effect dependent on both the pore fraction and the morphology of the pores (Ross, 1960; Mogard et al., 1964).

It is in the low-temperature region that irradiation has a noticeable effect on the conductivity of UO_2. There, the irradiation-induced defects in the lattice are not sufficiently mobile to be immediately annealed, while the lattice perodicity of the unirradiated material is large enough for damage to be apparent. The combined effects of irradiation exposure and annealing temperature have been most systematically studied by Ross (1960) and J. L. Daniel et al. (1963). Specimens irradiated by the latter at some temperature in the range 0 to 100°C showed appreciably lower conductivities than unirradiated specimens in room-temperature measurements (Fig. 3–7). When the post-irradiation measurements were extended to higher

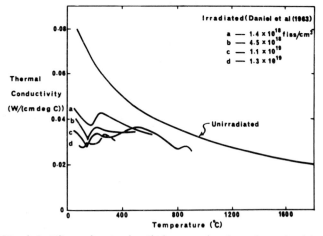

Fig. 3–7. Effect of *prior* irradiation on the thermal conductivity of sintered UO_2 as a function of the temperature of measurement. (After J. L. Daniel et al. (1963).)

temperatures, annealing stages were observed at about 200 and 400°C. Even after annealing at 600°C, some damage still remained in a specimen exposed to 1.1×10^{19} fissions/cm³ (0.05 % uranium burnup).

Ross measured the thermal conductivity at 60°C of specimens that had been previously irradiated to various exposures. He estimated his specimens to have been at some temperature below 500°C in the reactor, and it seems likely that they were at a higher temperature than those of Daniel et al.. In the range 5×10^{14} to 2×10^{16} fissions/cm³ the conductivity decreased with exposure: Thereafter, little, if any, change was apparent up to 6×10^{18} fissions/cm³. The results of Daniel et al. are in general agreement, but suggest that the damage for lower irradiation temperatures has not saturated by 1.34×10^{19} fissions/cm³.

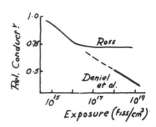

Ross's further measurements following anneals demonstrated the complex nature of the damage even in his specimens, which presumably did not contain the low-temperature damage. Figure 3–8 shows that an anneal

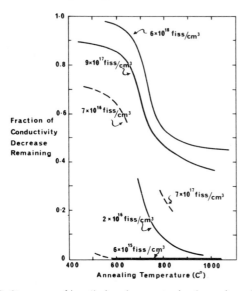

Fig. 3–8. Recovery of irradiation damage to the thermal conductivity at 60°C of sintered UO_2 as a function of the annealing temperature, for different exposures. (After Ross (1960).)

at relatively low temperatures, say 600°C, removes most of the damage from lightly irradiated specimens, but leaves the more heavily irradiated ones practically unaffected. The latter respond to higher temperatures, but the amount of recovery appears to saturate at about 50% over 900°C. The annealing stage seen about 400°C in lightly irradiated material moves to higher temperatures and divides into two components at higher exposures. Each stage is presumably associated with the annihilation or clustering of irradiation-induced lattice defects, but their identifications have not yet been established. For a burnup of well under one percent of the uranium in these specimens, the impurity effect of fission products is probably insignificant.

Measurements made with thermocouples in the UO_2 during irradiation generally show similar reductions in the thermal conductivity at low temperatures as the exposure increases. Bogaievski et al. (1963), Hawkings and Robertson (1963), and Hawkings and Bain (1963) all observed such reductions for temperatures below 500°C and for exposures up to 10^{19} fissions/cm^3 (Fig. 3–9). Clough (1962) found lower conductivities for UO_2 specimens during

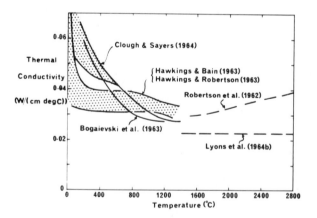

Fig. 3–9. The thermal conductivity of sintered UO_2 during irradiation as a function of the temperature. The solid lines were derived from thermocouple readings, the interrupted lines from interpretation of structures. In the shaded regions the values were found to depend on the exposure.

irradiation than for unirradiated ones but observed no progressive decrease in the same range of exposures. His specimens had not been submitted to irradiation at temperatures below 170°C, so he suggested that less damage may persist in specimens irradiated at an elevated temperature than in ones irradiated at a low temperature and subsequently heated to the higher temperature.

Cohen et al. (1961) concluded from their experiments that there was no decrease in the conductivity of UO_2 during irradiation, either initially or progressively, with exposure up to 4×10^{19} fissions/cm³: Rather, they attributed increases in the fuel temperature during successive reactor cycles to possible expansion of the sheath and cracking of the fuel, causing a deterioration in the coefficient of heat transfer between the fuel and the sheath. It has been argued (Robertson et al., 1962) that their results can be explained by supposing that the conductivity is damaged during irradiation at low temperatures, and that the damage is partly annealed during subsequent exposure at higher temperatures.

R. C. Daniel and Cohen (1964) have taken UO_2 to very high exposures and have reported a decrease of 50% in the conductivity of a specimen, whose average temperature was 525°C, at 2.5×10^{21} fissions/cm³. By this exposure, fission products and their oxides constitute an impurity content of about 20 mol%, so a reduction of this magnitude at relatively low temperatures is not surprising. Daniel and Cohen revised downwards the earlier calculations by Cohen et al. of the "effective conductivity", i.e., the overall conductivity of the fuel and the fuel/sheath interface. In both series of experiments, the "effective conductivity", even at low exposures, was appreciably lower than the true conductivity of unirradiated UO_2, but the difference was attributed to the temperature drop at the interface. Thus, the authors reaffirmed that the conductivity of sintered UO_2 is virtually unaffected by irradiation up to 10^{19} fissions/cm³ at temperatures around 500°C.

Stora et al. (1964) measured the thermal conductivity

of UO_2 while under irradiation to 10^{19} fissions/cm^3. In the range 500 to 1300°C, they observed no significant differences between conductivity values obtained in the reactor and comparable ones they obtained from a laboratory apparatus. Their results were in good agreement with those of their colleagues Bogaievski et al. Considerable weight must be given to these two determinations, since both employed thermocouples at different positions in the fuel and also in the sheath. Thus, these investigators were able to separate the temperature drops at the interface and in the fuel to obtain a true conductivity of the UO_2 without assumptions regarding the interfacial heat-transfer coefficient.

The situation has since been greatly clarified by a well-controlled experiment due to Clough and Sayers (1964). Not only were there sufficient thermcouples in their three capsules to separate the temperature drops at the interface and within the fuel, but also fuel temperatures were maintained constant throughout the irradiation. The results (Fig. 3–10) showed that the conductivity of UO_2 exposed at 180°C decreased 22% by 2×10^{18} fissions/cm^3, while by 4×10^{18} fissions/cm^3 that at 320°C decreased 17%, and at 520°C there was essentially no change. The

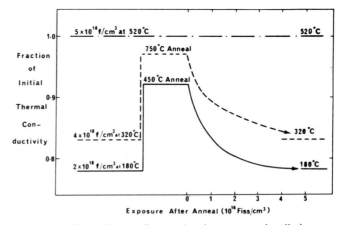

Fig. 3–10. Effect of annealing and subsequent re-irradiation on damage to the thermal conductivity of sintered UO_2: Measurements made during irradiation at the temperatures indicated. (After Clough & Sayers (1964).)

conductivity of the first two was partly restored by raising the temperature to 450 and 750°C, respectively. It is notable, however, that during further irradiation, damage occurred much more rapidly than it had initially. The conductivity of irradiated UO_2, therefore, depends on the thermal and irradiation history as well as the temperature and exposure at the time of measurement. Even short periods at low temperature during reactor startup are probably enough to damage the conductivity significantly.

In the temperature range 1250 to 2000°C there have been very few reliable measurements of temperature in UO_2 during irradiation, and the thermal conductivity has been largely inferred from observations of grain growth (Robertson et al., 1962; Lyons et al., 1964b). The conductivity is apparently not substantially affected by the irradiation, but the recent finding that irradiation impedes grain growth (Sec. 3.2) means that the method is valid only for light irradiations. More measurements are needed in this temperature range, using specimens with thermocouples in both fuel and sheath and taking the fuel to over one percent burnup of the uranium. However, since Clough and Sayers observed no effect of irradiation on the conductivity in the range 500 to 1600°C for exposures up to 4×10^{19} fissions/cm³, it is unlikely that any large change would be found. At higher temperatures, a mean value for the conductivity during irradiation can be deduced from the extent of melting in the UO_2 observed subsequently.

There is still much controversy concerning the thermal conductivity of sintered UO_2 that has undergone structural changes, both during irradiation and in its absence. J. L. Daniel et al. (1963) discovered that the conductivity, measured in the laboratory, of a single-crystal specimen did not decrease with increasing temperature as rapidly as that of normal polycrystalline material and even passed through a minimum about 700°C. At 1200°C, the single crystal had a conductivity greater by a factor over two. However, this was the highest temperature at which meassurements were made, and the natural scatter did not permit extrapolation with any great confidence. Further measurements on other specimens and in other apparatus have confirmed that the conductivity of single crystals around 1000°C can be two to three times the value for normal polycrystalline material but showed that the enhanced conductivity disappeared as the temperature was further increased (Pashos et al., 1964). Nevertheless, more accurate determinations in this area would be welcome.

May et al. (1962) studied UO_2 single crystals in a very simple comparator that could arrange specimens in the order of their thermal conductivities at elevated temperatures without providing any numerical values. They found that the composition had a significant effect with the higher conductivities in the hypostoichiometric specimens (UO_{2-x}, where x is about 0.01). The work was extended to demonstrate the same qualitative effect in polycrystalline sinters. Although early comparisons showed no difference between single crystals and polycrystalline sinters of the same composition, later work (May and Stoute, 1965) showed that the single crystals have a slightly higher conductivity. Using a quantitative comparator, Kollie et al. (1964) found the conductivity of hypostoichiometric material to be 17% greater than that of stoichiometric UO_2 in the range 75 to 300°C. Using a radial-flow method, Hetzler and Zebroski (1964) found an increase of approximately 50% for the hypostoichiometric material in the range 800 to 1400°C.

Among others, Christensen (1963b) has estimated possible sources of enhanced conductivity in UO_2. He calculated that heat transfer by radiant energy should be significant in stoichiometric UO_2 at temperatures between 500 and 1000°C. Any excess oxygen, by contributing free charge carriers that scatter the radiation, rapidly reduces the infrared component of heat transfer. On the assumption that hypostoichiometric UO_2 contains fewer free carriers than nominally stoichiometric material, the effects of composition on the conductivity are qualitatively explained. Unfortunately, the magnitude of the total conductivity calculated is much less than that measured for single crystals, and the observed maximum in the conductivity occurs a few hundred degrees higher than calculated. Christensen argued that an electronic contribution could become significant but only at temperatures over 1500°C. Such a conclusion, however, must be confined to hyperstoichiometric material until the electrical properties of hypostoichiometric UO_2 have been determined. In short, the existence of enhanced conductivity has been adequately demonstrated, but the source has not yet been identified.

De Halas and Horn (1963) argued that the metallographic stucture of certain irradiated UO_2 specimens was consistent with the large grains at the centre having the high conductivity of single crystals. Essential to their argument was the assumption that fuel which had initially melted (out to a radius 0.3 cm in Fig. 3–11) had subsequently been solid for the remainder of the irradiation. Hausner and Nelson (1965), however, displayed micrographs showing that a structure similar to that of Fig. 3–11 is modified during continued irradiation at slightly lower temperatures. Even if the specimens irradiated by de Halas and Horn had suffered a reduction in fuel temperatures during the irradiation, it would be necessary to demonstrate that an increase in thermal conductivity was responsible. Other causes, such as a decrease in neutron flux, burnup of fissile atoms, or sintering of the powdered UO_2 used in some of their specimens could all explain the effect.

To test the hypothesis more critically, Notley (1963) irradiated sintered UO_2 at a power sufficient to cause substantial grain growth. He then re-irradiated the same specimen at a higher power. From the extent of the molten region observed in post-irradiation examination, he deduced that the thermal conductivity of the large grains was little higher, if any, than that of the original UO_2, and certainly not as high as the value supposed by de Halas and Horn. Using thermocouples at different radial positions in sintered UO_2 pellets, Christensen and Allio (1965) showed

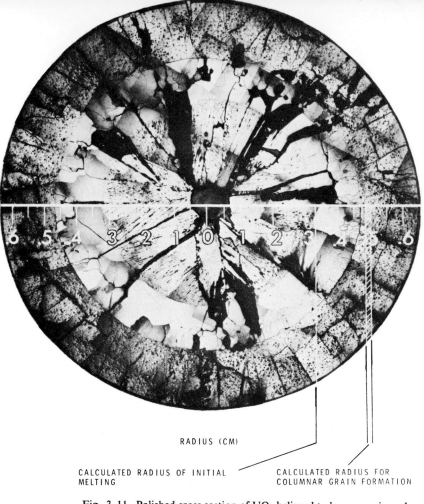

RADIUS (CM)

CALCULATED RADIUS OF INITIAL
MELTING

CALCULATED RADIUS FOR
COLUMNAR GRAIN FORMATION

Fig. 3–11. Polished cross section of UO_2 believed to have experienced
central melting during irradiation. (After de Halas & Horn (1963);
captions on photograph refer to the original text.)

that the thermal conductivity in the range 1700 to 2150°C
did increase as columnar grains developed but by only
about five percent. Such an increase could be explained
by the associated densification of the fuel. Notley and
MacEwan (1965) found that the effect of varying the
initial density of the sinter on the extent of grain growth
and on the release of fission-product gases was that expected
from increasing conductivity with decreasing porosity.

In another experiment, MacEwan et al. (1964) irradiated
samples of polycrystalline $UO_{1.98}$ alongside controls of
$UO_{2.007}$. Central thermocouples showed lower tem-

peratures for the hypostoichiometric material by as much as 150 deg C at 1700°C. Measurements of the electrical potential developed between the thermocouple sheath and the fuel sheath during irradiation indicated that the $UO_{1.98}$ was n-type above 500°C, while the $UO_{2.007}$ was p-type at all temperatures. These results suggested that irradiation may cause dissolution in the UO_2 lattice of some uranium that would normally be precipitated at temperatures below 1500°C, as suggested by Childs from studies of stored energy (Sec. 3–10). Thus, further determinations over a wider range of compositions seem desirable. Irradiations with central thermocouples in single crystals of both stoichiometric and hypostoichiometric UO_2 would also appear worthwhile, in view of the large effect seen in the earlier laboratory experiments, but not yet explained.

There has, of course, been speculation that the hot central region of the UO_2 would become hypostoichiometric during operation. A metallic sheath, the outer cool UO_2 and, possibly, the solid fission products are potential getters for some of the oxygen in the hottest UO_2. The observation of white, possibly metallic, precipitates in the region of structural change (Sec. 3.8) is supporting evidence.

From all the work performed on sintered UO_2, it appears that irradiation has an appreciable effect on its thermal conductivity only at temperatures below 500°C. Damage incurred below 500°C, however, persists during irradiation at higher temperatures. At temperatures over 500°C, thermal vibration of the atoms has already caused phonons to be scattered at nearly every lattice spacing, so irradiation damage can probably produce little additional disruption. Although some laboratory determinations have found the thermal conductivity of UO_2 to increase with increasing temperature above 2000°C, experiments under irradiation suggest that the increase is either small or absent altogether. Similarly, the increased conductivity obtained for single crystals in the laboratory has not been observed in the reactor. However, under irradiation, the increase has been looked for at temperatures higher than those at which it has been found in the laboratory.

The use of hypostoichiometric compositions appears to give some benefit, even under irradiation, but the reason is not understood.

It is hard to believe that the cracks seen in UO_2 sinters after they have been irradiated do not constitute serious thermal barriers. However, the cracks in the outer region are predominantly radial, and hence do not much affect the heat flow: Those in the inner, hot region form only on shutdown and heal during subsequent operation (Sec. 3.2). Even the circumferential cracks in the outer, non-plastic region, whether naturally occurring or deliberately simulated, do not significantly affect the extent of grain growth. From such comparisons, Robertson et al. (1962) and Bain (1963) have estimated minimum heat-transfer coefficients of 0.85 and 1.5 $W/(cm^2 \deg C)$ for cracks in fuel held tightly in its sheath. Where large fuel/sheath diametral clearances permit the cracked fragments to remain loose, even when thermally expanded, significant temperature drops across the cracks may be expected.

In April 1965, the International Atomic Energy Agency convened a panel to discuss the thermal conductivity of UO_2. When the report is published it should constitute the most authoritative review of the subject.

3.5 Thermal Expansion

As with some other aspects of the behaviour of UO_2 under irradiation, thermal expansion is determined largely by the temperature distribution and is little affected by irradiation *per se*. The expansion is important in practice because of the very high temperatures prevalent in UO_2 fuel elements. There is a general impression that the expansions of elements containing cracked UO_2 are erratic, and that such a situation is inevitable. This pessimism is partly due to the fact that experimental difficulties long prevented direct or accurate measurements of the expansions during irradiation. Conclusions drawn from post-irradiation measurements depended on reliable knowledge of how the power output had varied throughout the irradiation and how the sheath dimensions changed on final shutdown. Also, the relevant material properties of unirradiated UO_2, the coefficient of thermal expansion over the full temperature range, and the volume change on melting were not reliably established. Despite these difficulties, much is already known about the factors governing expansions of UO_2 fuel elements.

The longitudinal expansion of sheathed UO_2 was measured continuously during irradiation by Notley and Harvey (1963). They demonstrated that the elements elongated when first brought to power, with the amount of elongation greater for higher fuel temperatures, but they found that the elements then slowly contracted a little at constant power. The contraction was attributed to plastic deformation of the UO_2 under the longitudinal load exerted by the sheath and the pressurized coolant. Plasticity of the hot UO_2 had previously been observed when preformed cavities were filled during irradiation (Robertson et al., 1962). Quantitative analysis of the results indicated that sheath elongation was determined by the expansion of UO_2 at about 750°C, with fuel at higher temperatures behaving plastically under irradiation. Since laboratory experiments by Armstrong et al. (1962) would not have predicted appreciable plasticity below 1200°C, irradiation-induced creep at relatively low temperatures is suggested. If so, the rate of fissioning may have an effect. From comparisons of the elongation of specimens irradiated in different experiments, Notley and Fitzsimmons (1962) concluded that increasing the sheath strength decreased the fuel elongation, by causing appreciable creep in the fuel to occur at lower temperatures. Similarly, heating the fuel more rapidly raised the temperature of the onset of appreciable plasticity thereby increasing the elongation.

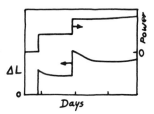

In practical fuel elements, advantage is taken of the plasticity of the UO_2 by preforming cavities in the fuel pellets to allow for expansion. In the end face of each pellet, there is dishing that is surrounded by a narrow shoulder in contact with the adjacent pellet. The elongation of the stack is then controlled by the expansion of the shell of UO_2 immediately outside the dishing (Notley and Harvey, 1963).

Movement of the expanding stack into any clearance left at the end of the element is determined by the frictional forces between the fuel and the sheath. Using maximum displacement markers inside the sheath at the end of the stack, Notley (1962) showed that the fuel elongated readily when assembled in free-standing sheaths with generous diametral clearances. However, when the fuel and sheath were in tight contact over their cylindrical surfaces, because the pressurized coolant had collapsed a thin sheath or due to small initial diametral clearances, the markers had scarcely moved. The small movements measured were consistent with estimated values (Notley et al., 1964).

A further source of longitudinal clearance is usually present, distributed along the stack of pellets. Since the end face of a pellet is never absolutely perpendicular to the pellet axis, there are wedge-shaped gaps between adjacent pellets. Although the gaps are thin, they can represent something of the order of one percent of the fuel length. When the pellets fragment under thermal stresses, this distributed clearance is available to accommodate fuel expansion. The remaining fuel elongation produces sheath elongation: The elastic component of the sheath's deformation disappears on reactor shutdown.

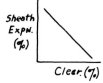

The dependence of the sheath's diametral strain on the diametral fuel/sheath clearance is simpler. For a given temperature distribution in the fuel, an increase in the clearance resulted in a decrease in the residual strain by the same amount, until the latter became zero (Bain et al., 1961). Figure 3–12 illustrates the fuel's diametral expansion (calculated from the residual strain of the sheath) as a function of its calculated centre temperature for a variety of specimens; all with fuel-surface temperatures estimated to be $400 \pm 100°C$. The increase in the slope of the band of experimental points near 2800°C is attributed to the

onset of central melting with its consequent volume
change of approximately 10%. Some of the scatter of
experimental points represented in Fig. 3–12 may be due

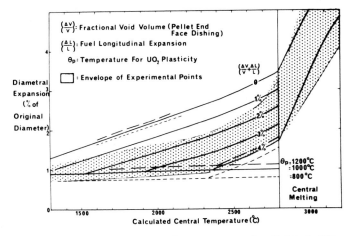

$\left\{\frac{\Delta V}{V}\right\}$: Fractional Void Volume (Pellet End Face Dishing)

$\left\{\frac{\Delta L}{L}\right\}$: Fuel Longitudinal Expansion

θ_p: Temperature For UO_2 Plasticity

☐ : Envelope of Experimental Points

Fig. 3–12. The diametral expansion at power of cylindrical UO_2
pellets as a function of their centre temperature for a constant surface
temperature of 400°C. The lines are calculated from the model
discussed in the text while the shaded band is the envelope of experimental observations. (After Notley et al. (1964).)

to the differences of UO_2 density, sheath material, thickness
and temperature, rate of startup, and duration of irradiation
for the specimens included in this single graph.

The UO_2 pellets can be idealized as an outer annulus
of elastic fragments containing an inner core of plastic
material as suggested by the elongations (Notley et al.,
1964). The interface between the two regions, of radius R_p
and at temperature θ_p, defines the points of contact between
adjacent fragments. Where the fractional free volume
$\Delta V_0/V$, due to porosity and cavities in the plastic core, is
sufficient to accommodate volume changes of the fuel
within that zone, the pellet's radial expansion ΔR is simply
that of the outer annulus. The latter equals the radial
expansion of the elemental shell at R_p plus the radial
expansion of an individual fragment. Thus,

$$\Delta R = R_p \alpha \theta_p + (R_s - R_p)\, \alpha \left(\frac{\theta_s + \theta_p}{2}\right), \qquad (3\text{–}2)$$

where R_s is the pellet's radius, θ_s its surface temperature, and α the linear coefficient of thermal expansion of UO_2. (The second term in Eq. (3–2) ignores curvature of the fragment but is an adequate approximation when $R_p > \frac{1}{2}Rs$.)

Where the free volume in the fuel is insufficient, an isostatic pressure in the plastic core forces the fragments outwards by an additional amount. To evaluate the radial movement of a point at R_p on heating from 0°C, the corresponding volume increase of the plastic core is calculated in two different ways

$$\int_0^{R_p} 2\pi RL3\alpha\theta \, dR - \pi R_p^2 L \frac{\Delta V_0}{V} + \pi R_m^2 L \frac{\Delta V_m}{V}$$

$$(3\text{–}3)$$

$$= \pi R_p^2 L \frac{\Delta L}{L} + 2\pi R_p L \Delta R_p,$$

where $\Delta L/L$ is the fractional elongation of the stack of pellets, $\Delta V_m/V$ the fractional volume change on melting, and R_m the radius of the molten region. Hence, the radial movement at R_p is given by

$$\Delta R_p = \frac{R_p}{2} \left[\frac{6}{R_p^2} \int_0^{R_p} \alpha\theta R \, dR - \left(\frac{\Delta V_0}{V} + \frac{\Delta L}{L} \right) \right.$$

$$(3\text{–}4)$$

$$\left. + \left(\frac{R_m}{R_p} \right)^2 \frac{\Delta V_m}{V} \right].$$

Then the total radial expansion of the UO_2 pellet is either

$$R_{\text{total}} = \frac{1}{2}(R_s - R_p)\alpha(\theta_s + \theta_p) + R_p\alpha\theta_p$$

or

$$(3\text{–}5)$$

$$R_{\text{total}} = \frac{1}{2}(R_s - R_p)\alpha(\theta_s + \theta_p) + \Delta R_p,$$

whichever is the greater. Equation (3–5) is plotted in Fig. 3–12 for a few values of the parameters. In practice, preformed cavities and porosity in the UO_2 have been shown to decrease the radial expansion (Notley and MacEwan, 1965), while the rate of bringing the fuel up to temperature (which affects θ_p) may also be important.

After irradiation, a fuel rod containing UO_2 pellets often looks something like a bamboo cane, because of the circumferential ridges that developed in the sheath at pellet interfaces. This observation stresses the discontinuous nature of the stack of pellets, a feature that was ignored in the idealized model. Stress analysis of an elastic, finite cylinder subjected to a parabolic temperature distribution shows that the diametral expansion of a pellet increases towards its ends. Thus, the ridges could be explained qualitatively if it were supposed that under irradiation UO_2 sinters do not crack until they achieve a temperature differential much greater than that predicted from the properties of unirradiated material (Sec. 3.2). Otherwise, the pellet must expand as a coherent whole even when severely cracked. No detailed and fully satisfactory theory is available to explain the formation of these ridges, but their existence serves as a useful reminder of the approximations involved in any model for such a complex system.

Bain et al. (1964c) succeeded in extracting several whole pellets that had been exposed to very high centre temperatures in an irradiation consisting of a single, short power cycle. The pellets were extensively cracked, but the individual fragments were adhering to the central core which, in many instances, had been molten. Initially, these specimens had fuel/sheath diametral clearances equal to a few percent of the diameter and, after irradiation, the pellets were found to have increased appreciably in diameter. For a given power per unit length of fuel, the residual expansion of the UO_2 increased with the original clearance. Thus, even without the fuel swelling (Sec. 3.7), UO_2 under irradiation tends to fill voids available to it either by plastic flow of the central region or outwards movement of the fragments.

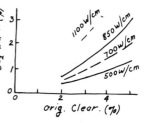

3.6 Fission-Product Gas Release

Swelling due to the retention of fission-product gases in UO_2 has been encountered (Sec. 3.7), but it is the release of these gases from UO_2 that has caused more concern so far. Since gas released from the fuel may produce a net internal pressure on the sheath, factors affecting the release have been extensively investigated. About 15% of the fission-product atoms are the rare gases krypton and xenon so, very roughly, a piece of UO_2 generates its own volume of gas, measured at standard temperature and pressure, at 0.4% burnup of the uranium (10^{20} fissions/cm^3). Much, therefore, depends on what fraction of the total generated is released, and many empirical results have been reported for specific designs of fuel that operated under arbitrary, and usually variable, conditions. Experiments that elucidate the mechanisms of release are less plentiful.

The stable and long-lived isotopes released during irradiation can be collected and analysed afterwards by puncturing the sealed fuel element in an evacuated enclosure. Such tests, when conducted on specimens made from a uniform batch of UO_2 and irradiated together, have shown clearly that the fractional release increases with increasing centre temperature of the fuel (Robertson, 1963). Lewis et al. (1964), by analysing for the gas remaining in the fuel as a function of the sample's radial position within the sintered pellet, showed that little is released below 1000°C and that little remains in the UO_2 above 1800°C. Even more direct proof was obtained by Melehan et al. (1963) who measured the continuous release during irradiation from small specimens of fused or sintered UO_2, as a function of their temperature, by following the activity of relatively short-lived isotopes. The results, interpreted in terms of a diffusion coefficient, are reproduced in Fig. 3–13. In the temperature range from 1000 to 1650°C, the release was markedly temperature-sensitive: At lower temperatures, the release was practically constant. The latter observation confirmed a similar one by Markowitz

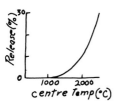

(1957) and has since been reconfirmed by Carroll and Sisman (1965). No difference was found in the behaviour of kryton and xenon, after allowance had been made for their fission yields and decay constants.

The small and nearly constant release at low temperatures is commonly said to represent escape by recoil. A fission fragment is produced with an energy high enough to give it a range of several microns through the UO_2 lattice. Thus an appreciable fraction of the fission products generated in a very thin surface layer recoil out of the UO_2. From a knowledge of the range and the specific surface area of the UO_2, the fractional release from the whole specimen can be readily calculated, and the result often agrees well with the observed value. However, the fission fragment, in general, escapes with considerable energy that is diminished little by passing through a thin layer of gas. It should, therefore, penetrate the next solid in its path, either another UO_2 surface or the sheath, and not contribute to the net release. For a recoil mechanism, the fractional release rate should be the same for all isotopes of a given element, whereas Melehan et al. (1963), Carroll (1963), and Schurenkamper and Soulhier (1964) all found the rate depended on the decay constant of the isotope. This observation suggested that the gas had been delayed for a finite period before being released. Except for fuel in a liquid-metal heat-transfer medium, release by recoil is probably negligible.

Experiments by Bierlein and Mastel (1962) provided another explanation for the observed releases from fuel at low temperatures. They found that the energy released by fission fragments in thin films of UO_2 was sufficient to eject significant amounts of material that collected on an adjacent surface. Rogers and Adam (1962a) had demonstrated a similar effect with uranium metal, even when the surface was oxidized. Rogers (1964, 1965a) showed that a dynamic equilibrium is soon established between ejection from the original uranium surface and re-ejection from the deposit, which consists of a series of discrete spots. Later, while the amount of deposited material remains constant, the number of spots decreases and a thin film forms between them. Rogers interpreted his results to show that a particle containing about 10^4 atoms was ejected from uranium by each fission spike intersecting the surface, and that a spike occurred every 1200 Å along a fission-fragment track. To explain the early saturation in the amount of uranium ejected, Rogers supposed that a spot of deposited uranium was vapourized by ionization when a fission fragment passed through it.

Transmission electron microscopy (Sec. 3.2) shows that in fine-grain UO_2, such as that on the surface of uranium or in most thin films, the energy release is concentrated around the fission fragment's track: Therefore, less material might be expected to be ejected from coarse-grain material. Experimentally, Rogers (1965b) demonstrated that the amount ejected per fission from a thin film of UO_2 decreased drastically as the grain size increased from 50 to 180 Å during the exposure. Nilsson (1964), who is, unfortunately, the only one to have studied the phenomenon on bulk sintered UO_2, found that only nine uranium atoms were ejected per fission fragment, i.e., no more than would be expected from the emission of knock-ons. Thus, in practice, the depth of UO_2 at free surfaces contributing to gas release by the ejection mechanism is probably a small fraction of the range of fission fragments in the fuel. The contribution, therefore, is unlikely to be as much as that originally calculated for release by recoil.

During the post-irradiation annealing of UC, Auskern (1964) found the fission-product gas was released with an unusually low and variable activation energy in the range 150 to 975°C (Sec. 4.5). He ascribed the result to relatively easy diffusion within the first few unit cells near the surface. Since a nearly similar low-temperature release from UO_2 has been observed in post-irradiation annealing experiments reported by Belle (1961), the same phenomenon may be operative in both fuels.

In the temperature range of 1000 to 1650°C, volume diffusion has been generally regarded as the predominant mechanism for gas release. Results from continuous measurement of release during irradiation showed the release to be a thermally activated process (Fig. 3–13), but the bulk of the evidence comes from post-irradiation anneals. Samples consist of UO_2 sinters or single crystals, sometimes crushed and sieved to yield particles of uniform size: The specific surface area is measured by gas adsorption techniques. These samples are irradiated at low tempera-

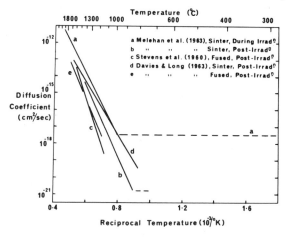

Fig. 3–13. Dependence on temperature of the diffusion coefficient for fission-product gases in UO_2.

tures to produce a uniform distribution of fission products whose release can be followed by counting their radio-active emission, when the sample is subsequently heated. The dependence of the release of rare gas from the particles on their size, the temperature and the duration of the anneal, was that expected of a diffusion-controlled process (Booth and Rymer, 1958).

If the particles have a simple shape, the diffusion equation can be solved to yield the diffusion coefficient for the particular rare-gas isotope. An approximate solution, valid for fractional releases less than 0.1, is the same for a variety of shapes, e.g., sphere, cylinder, and cube

$$f = 2(S/V)(\sqrt{D/\pi})\sqrt{t}, \qquad (3\text{–}6)$$

where f is the fractional release in time t from a body of surface-to-volume ratio S/V and D is the diffusion coefficient. This independence of shape reflects the fact that a small release is drawn exclusively from a thin surface layer: The radius of curvature of the surface is therefore unimportant as long as it is large compared with the layer's thickness. A less fortunate consequence is that the diffusion coefficient obtained is that applicable to a thin surface layer which may or may not be representative of the bulk. A more exact solution is available for analysing large releases (Lewis, 1960).

Values for the diffusion coefficient of xenon in UO_2, deduced from post-irradiation anneals by the method of the previous paragraph, are shown in Fig. 3–13. The activation energy is high; in the neighbourhood of that for the self-diffusion of the uranium atoms. Long et al. (1964) argued from the results of some of their experiments that the xenon cannot diffuse by single vacancies in the oxygen sublattice: But Matzke (1965) has drawn a similar conclusion with regard to the uranium sublattice from other experiments, so diffusion via interstitial positions or by gas-vacancy clusters is being considered. Results obtained by Davies and Long (1963), shown as lines d and e in Fig. 3–13, suggest that the diffusion coefficient for xenon is different in sintered and fused UO_2. However, other results do not support this suggestion. It appears more likely that the different behaviours are due to minor differences in composition: The plasma-fused specimens received a vacuum anneal at 1650°C that the sinters did not have.

Using a series of sinters covering a wide range of surface-to-volume ratios, Long et al. (1960) established the relation

$$\log(f^2/t) = \text{constant} + 2\log(S/V) \qquad (3\text{–}7)$$

for small releases at a constant temperature. Thus, as long as the gas is only coming from a thin layer at the free surface, Eq. (3–6) applies to sinters, too. As a mathematical convenience in solving the diffusion equation, the sinter is often replaced by an idealized model of "hypothetical sphere." The specimen is imagined as an assemblage of uniform spheres having the same surface-to-volume ratio as the sinter. The radius a of any sphere is therefore given by

$$3/a = S/V \qquad (3\text{--}8)$$

and

$$f = (6/\sqrt{\pi})(\sqrt{D/a^2})\sqrt{t}. \qquad (3\text{--}9)$$

The reason that the concept has proved fruitful is that D and a remain together in the combination D/a^2 throughout the mathematical treatment. The product, called the composite diffusion coefficient or D', serves to characterize the sinter. If the activation energy for diffusion is known, an experimental measurement of the slope of a curve of f versus $\sqrt{t}$ at one temperature (usually 1400°C) permits calculation of the release from that batch for any other annealing time and temperature. For small releases, replacing the actual sinter by the hypothetical model does not in itself introduce significant errors; but for releases over 50%, say, the predictions are only approximate, even when the model is applied to the more exact version of Eq. (3–6).

At a particular temperature, D has a specific value representative of UO_2, while a is characteristic of the sinter and is related to its value of (S/V) by Eq. (3–8). Since Belle and Lustman (1957) showed that the surface-to-volume ratio is related to the sintered density, it was to be expected that the composite diffusion coefficient, too, depends on the density. Stevens et al. (1960) demonstrated just such a correlation, with a high value of D' at low densities dropping off rapidly with increasing density over 10 g/cm³, where the interconnected open porosity also drops off rapidly. The empirically established dependence can be useful for predicting approximate values of D', but Stevens et al. showed that altering the fabrication route could significantly change the correlation. Where the porosity had become sealed off early in the sintering,

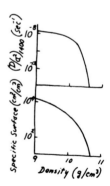

the exposed surface area and, hence, the value of D', fell off at a lower density.

Thus far, the simpler aspects of gas release have been stressed, but the rest of the section is largely devoted to other aspects that make the overall process a most complicated one. Even for small releases during post-irradiation annealing, departures from the ideal behaviour of a simple diffusion model are experienced under certain circumstances.

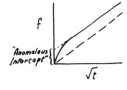

An initial burst of gas (the "anomalous intercept" on a plot of f versus $\sqrt{t}$) results from even slight oxidation before or during the diffusion anneal (Stevens et al., 1960). The xenon release from three sinters of different composition during isochronal annealing under reducing conditions is shown in Fig. 3–14 (Lewis et al., 1964). Below 1300°C, release from the hyperstoichiometric specimens is more rapid than from the nominally stoichiometric one, in accord with earlier results by Lindner and Matzke (1959) in the same relatively low-temperature region. At 1300 and 1400°C, the increased release from the hyperstoichiometric specimens was accelerated, but, above 1400°C, all three behaved similarly. Analyses of $UO_{2 \cdot 03}$ exposed to the same treatment demonstrated that all the excess oxygen was evolved between 1100 and 1400°C. The authors, therefore, suggested that the evolution of excess oxygen, in addition to the mere presence of excess oxygen, could accelerate the release of xenon.

Bursts of gas released on cooling specimens from temperatures below 1000°C may also be attributable to slight oxidation, since they have not been observed in vacuum or under reducing conditions. Rothwell (1962) found that specimens that had been reduced to hypostoichiometric compositions during a diffusion anneal around 2000°C subsequently released a burst on cooling, presumably on transformation to UO_2 and uranium metal. Lagerwall and Schmeling (1964) have discussed in detail several possible reasons for nonideal releases. Departures from linearity in the Arrhenius plot of the logarithm of the diffusion constant versus reciprocal temperature, and low values for the activation energy, may be due to very small amounts of

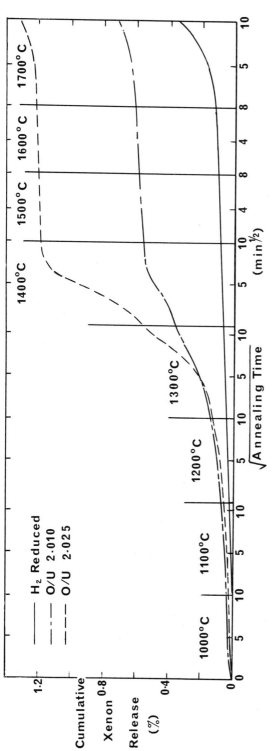

Fig. 3-14. The effect of initial stoichiometry on the release of xenon from sintered UO₂ during isochronal annealing under reducing conditions. (After Lewis et al. (1964), courtesy United Nations.)

impurity in the UO_2, but, at present, not enough is known of these effects on diffusion in compounds.

Application of an external stress, causing appreciable deformation, accelerates the release during post-irradiation annealing. Morgan et al. (1965) compressed hollow cylinders of sintered UO_2 and ThO_2–UO_2 that were being annealed between 1000 and 1550°C. The increase in release was greater for larger loads and at higher temperatures: It decreased steadily as the specimen was held at constant temperature with the load applied. The effect is probably associated with dislocations, but the mechanism is not established. Dislocations may act as paths of easy diffusion or, in moving, they may sweep out the gas. Even the accelerated release was less than 0.1% of the total gas, so localized effects at the point of load application cannot be excluded. Long et al. (1964), looking for the same accelerated release, concluded that there was no significant effect, but the deformation of their specimen was probably less.

Probably the most serious oversimplification in the model of "hypothetical spheres" is that it completely ignores the internal trapping of the gas. Closed pores are observed by optical microscopy in unirradiated UO_2 sinters and even in fused UO_2 be electron microscopy. MacEwan and Stevens (1964) collected gas released during pulverization of irradiated samples: The increased amount from specimens that had received an anneal between irradiation and pulverization represented the gas that had diffused to internal pores and become trapped. They also showed that the apparent values of the diffusion coefficient fell off drastically at high exposures, an observation completely unpredicted by the simple model. Arguing that the actual diffusion coefficient was unlikely to have been affected so markedly by irradiation, they concluded that trapping of the gas was responsible and fitted their results to a more sophisticated model analysed by Hurst (1962). Their detailed interpretation indicated that irradiation produced new pores, of under 100-Å diam, which have since been observed by electron microscopy (Sec. 3.2).

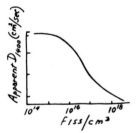

Thus, even if the original specimen were fully dense, trapping would still occur. The same post-irradiation annealing experiments suggested that the trapping was not permanent: This implies either appreciable solubility of the rare gas in the solid or migration of the pores themselves.

Following their observations of pore migration in thin metal films (Sec. 2.5), Barnes and Mazey (1964) looked for the same phenomenon in UO_2. They injected helium ions into thin films of oxidized uranium, then they developed pores by pulse annealing the specimens in the beam of an electron microscope. Further pulse anneals caused pore migration, with some pores erupting from a free surface and others coalescing. When coalescence occurred, the sum of the squares of the pore radii remained constant, just as in the metal films, which implies that the gas pressure was in equilibrium with the surface tension at the annealing temperature. Williamson and Cornell (1964), who performed a similar experiment with krypton in UO_2, found an inverse correlation between the pore radius and its velocity of migration and supposed that surface diffusion was the principal mechanism for migration. Since none of the pores was seen to shrink in size as a result of the annealing, there is no evidence from these short-term experiments for significant solubility of rare gases in UO_2.

A finite solubility of helium in UO_2 has been measured (Rufeh, 1964), but no comparable results are available for krypton and xenon with their larger atoms. Under irradiation, there exists another mechanism for the krypton and xenon to re-enter the UO_2 from the gas phase. Fission fragments recoiling out of the surface can, in collision with a rare-gas atom, transfer sufficient energy for the latter to be driven back into the solid. Where the fuel is at a high temperature, much of the gas that re-enters would soon diffuse out again, while gas driven into cool fuel would be subject to release by the low-temperature mechanisms already discussed. Chemical analyses of UO_2 that was irradiated in an atmosphere of natural xenon under high pressure have proved the existence of the phenomenon and provided an approximate evaluation (Lewis et al., 1964).

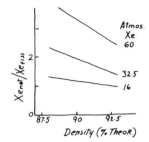

Over three times as much natural xenon as fission xenon was found in some samples of the fuel. For fuel temperatures under 1000°C, the amount of gas driven in was greater for lower densities (and, therefore, presumably higher specific surface areas) of the UO_2, for greater exposures (fissions/cm³), and for higher concentrations of xenon in the gas phase: At temperatures over 1500°C, the amounts of both natural and fission xenon retained by the fuel were much less, and their ratio more variable.

Lewis (1960) had earlier analysed such results to show that a pair of fission fragments, if they dissipated all their energy in the gas phase, would cause roughly a hundred xenon atoms to enter the UO_2. However, he pointed out that at high gas pressures some of the impelled gas atoms would lose their acquired energy in collisions with other atoms before striking the UO_2 surface. Thus, the amount re-entered would not be expected to increase linearly with increasing gas pressure. Still unresolved is the extent to which the amount of re-entered gas saturates if the available sites in a surface layer become occupied. Re-entry may, under certain circumstances, significantly reduce the internal pressure in practical fuel elements, but it could prove more important in the internal closed pores where the pressure can be very high.

In the same series of experiments, isotopic disproportionation of the xenon was observed (Lewis et al., 1964). In one element that was irradiated above the flux centre line in the reactor, UO_2 at the lower, and therefore hotter, end was relatively depleted in the light xenon isotopes but enriched in the heavy ones. In another element that had also been above the centre line but in much less of a flux gradient, the correlation was less clear. The authors supposed that the separation was effected while the gas atoms, probably ionized, migrated through the network of interconnected porosity. Migration by several different mechanisms and under the action of compounded driving forces seemed to provide the basic conditions for isotopic separation, but no detailed explanation was advanced.

Re-entry would have served to fix the gas atoms in positions where they were subsequently detected.

At temperatures over 1600°C, additional factors become relevant to the release of fission-product gases, especially in the thermal gradients occurring in fuel elements. Since this is the temperature range in which the original sintering was performed, prolonged exposure should cause further densification. If the amount of open porosity is diminished, the distance an atom has to diffuse before escaping from the solid would increase and the fractional release would be expected to decrease. In contrast, cracking and pore migration probably increase the gas release. Lenticular pores migrating up the thermal gradient (Sec. 3.2) are likely to carry along any rare gas in the UO_2 into a central void.

The normal isothermal growth of equiaxed grains also occurs in this temperature range. It has long been proposed, without evidence, that the moving grain boundaries would sweep out the gas, and thus accelerate release. Some results suggest that gas-filled pores hinder grain growth, but electron microscopy indicates that, when the boundaries eventually move, they do sweep fine pores along with them (Sec. 3.2). Thermal diffusion (as opposed to thermally activated diffusion in a concentration gradient) also may be important, but the question has not been investigated. Although unimportant in fuel elements, material loss by volatilization at temperatures over 1800°C has impeded laboratory studies by causing enhanced gas releases that may be attributed to grain-boundary sweeping: Mathematical techniques that correct for volatilization are now available (Lagerwall and Schmeling, 1964).

The experimental evidence indicates that gaseous fission products are captured by pores very soon after fission occurs. Thus, the gas release should be largely determined by the various forms of pore migration and growth discussed in Sec. 2.5. At temperatures sufficiently high for appreciable mobility of the pores, interaction with the grain boundaries becomes important. It is possible that gas release via interconnected porosity at the grain boundaries may then become significant. When a very high burnup is reached, pores throughout the UO_2 can become interconnected, releasing most of the gas (Sec. 3.7). Even at much lower burnup, most of the gas is released at temperatures over 1800°C, with the migration of large pores, over roughly 1μm in diameter, probably contributing to the release.

Measurements reported for the release of fission-product gas from sintered UO_2 have varied over several orders of magnitude, and, when they are all considered together, there is only poor correlation with any particular variable. This confused situation is largely attributable to imperfect control and knowledge of the many factors that can affect the release. Only when most of the potential variables were held constant, have clear correlations been established. With other materials, e.g., UC (Sec. 4.5), there is just as big a variation in the empirical results. Some limited success has been obtained in predicting gas releases by applying diffusion theory to a simplified model (Robertson et al., 1958; Belle, 1961). However, the model is so unreal in ignoring gas trapping and re-entry, grain boundaries, pore mobility, and any changes in the fuel during the irradiation that little significance can be given to it.

MacEwan and Lewis (1965) recognized that the hot central region, which is responsible for releasing most of the gas, bears little resemblance to either its initial condition or the ideal model. Therefore, they propose abandoning, rather than modifying, the simple model. Experimentally, the temperature distribution in the fuel has been shown to be most important, and its effect can be attributed to a thermally activated release without specifying any mechanism. The initial density has some correlation with the gas release, if only because any variations in density persist in the outer cool regions and, thus, by altering the thermal conductivity, affect the temperature distribution. The effect of fuel burnup on the *percentage* release was found to be small, so it could be expressed by a factor derived from experimental correlations. Small variations in the fuel's stoichiometry, the presence of minor impurities, and errors in estimating the temperatures could explain most of the remaining inconsistencies in their experimental results. Consideration of Secs. 3.2 and 3.4 suggests that the combined thermal and irradiation history of the specimen may be very important.

Interest in the release of fission-product gases from UO_2 stems principally from concern for the integrity of the sheath. Although the *amount* of gas released is important, it is the pressure exerted by the gas that could cause sheath deformations. In elements assembled with small clearances and high-density fuel, the free volume accessible to the gas during operation is typically about 1% of the fuel volume. Since the value is calculated as the initial free volume less the difference in expansion of fuel and sheath, consideration of Sec. 3.5 shows that there must be considerable uncertainty in the pressure deduced. Direct measurements of the pressure during irradiation have shown how the operating pressure, in general, increases with increasing burnup, and with an increase in the fuel temperature (Reynolds, 1963; Lewis et al., 1964). However, several irregularities have been observed: For instance, increases in the gas pressure tend to occur at, or immediately after, a reactor shutdown. More extensive studies are required, using the techniques now available for continuous measurement of gas pressure and fuel temperature.

3.7 Swelling

Substantial volume increases, associated with uniformly distributed porosity, can occur in irradiated UO_2 as well as in metallic fuels (Sec. 2.5). This swelling in UO_2, however, has only been observed after a high burnup, 5×10^{20} fissions/cm^3 and above: The type of swelling attributed to irradiation growth of uranium metal is not seen in isotropic UO_2. Barney and Wemple (1958) first noted the phenomenon both in a thin annular layer of UO_2 and in UO_2 particles from a cermet. The shape of the pores (Fig. 5–3 in Sec. 5.2) indicated appreciable plastic flow in the fuel, which by then contained approximately 10% of the uranium as fission products. The same authors showed that 0.5 wt$\%$ TiO_2 in the UO_2 caused markedly nonuniform porosity, but it was not clear whether the overall swelling was increased or decreased.

A systematic investigation of swelling in sintered UO_2 has since been performed by Bleiberg and his colleagues. Their experiments have greatly clarified the problem, but there still remain areas of ambiguity. In particular, most of their relevant specimens were flat plates containing fuel compartments, and there is some doubt as to how the swelling behaviour depends on the geometry.

In Fig. 3–15 are the cross sections of two fuel compartments after irradiation. The fuel of the first has bulged the sheath, with fuel and sheath still in contact. Such an appearance is representative of most of the specimens examined by R. C. Daniel, Bleiberg et al. (1962). Since the other dimensions remained almost unaltered, the volume change in the fuel could be deduced from measurements of the plate thickness. Determinations of the displacement density of the fuel from a few of the specimens confirmed the estimates. In the other compartment shown in Fig. 3–15, the sheath is bulged much further and is no longer in contact with the fuel, which is grossly swollen and contains many large pores. Although this compartment was intact when examined, such an appearance was rare, presumably because the later stages of sheath bulging proceed rapidly and soon lead to failure. Thus, while the early stages of sheath bulging truly reflect the fuel swelling, at some later stage it is the gas released from the fuel that determines the compartment's behaviour.

The swelling was followed as a function of the burnup by measuring the plate thickness at intermediate examinations during the irradiation. For any particular plate, the thickness increased slowly and continuously until a critical burnup. Thereafter, some of the compartments showed accelerated thickening and even sheath ruptures. Superposition of the curves for various plates (Fig. 3–16) shows remarkable similarity in their behaviour during the first stage, but the critical burnup at which the breakaway occurs depends on several factors. The fission-product gas released from the fuel and retained within the compartment was collected from enough specimens to establish the dependence of the release on the burnup (Fig. 3–17). Again, all the plates exhibit generally similar behaviour below the critical burnup at which the gas release increases sharply. Gas-release values could be obtained from only the intact compartments of those that had bulged seriously, but the few results available demonstrated that large increases in thickness and high gas releases coincided.

Daniel et al. suggested that the plot of volume change in the fuel as a function of burnup, prior to breakaway, should be considered as two intersecting straight lines. If so, the initial region of slow net swelling might be due to porosity in the as-sintered material accommodating some of the volume increase. However, the evidence is not thoroughly convincing because of the scatter in the measurements, and the results could also fit a power law of the sort attributed to the swelling of uranium metal (Sec. 2.5).

The fuel temperature is probably an important factor in the swelling of UO_2. The same authors showed that thermally isolated fragments of UO_2 contained gross porosity, while adjacent ones, cracked from the same pellet, were apparently free of pores if they were in good contact with the sheath. Something of this nature can be seen in Fig. 3–15b. The ends of the pellet are more effectively cooled and, hence, are less swollen than the central section.

Estimating the fuel/sheath heat-transfer coefficient, and hence the fuel-surface temperature, for any of these specimens is difficult since the compartments were evacuated during assembly. However, the variation between different specimens in the temperature drop across the interface is probably small compared with the variation between specimens in the temperature drop through the fuel. For comparison, therefore, specimens can be characterized by the integrated conductivity between the limits of surface and centre temperatures, calculated from the power out-

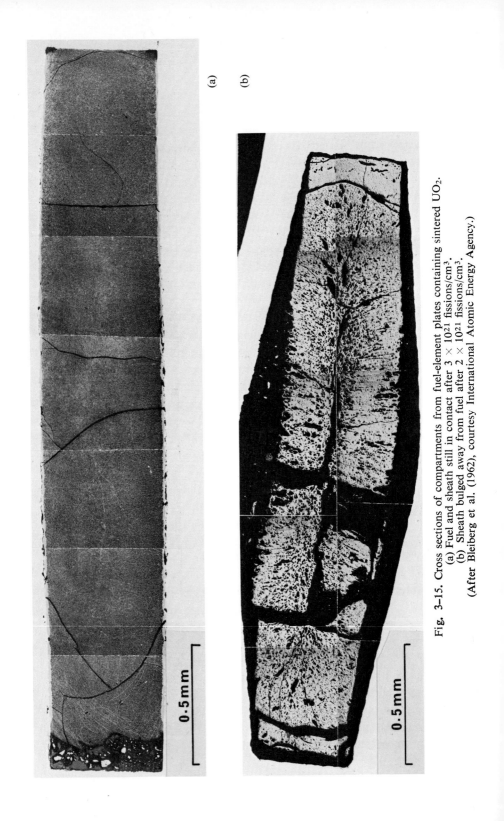

Fig. 3–15. Cross sections of compartments from fuel-element plates containing sintered UO_2.
(a) Fuel and sheath still in contact after 3×10^{21} fissions/cm³.
(b) Sheath bulged away from fuel after 2×10^{21} fissions/cm³.
(After Bleiberg et al. (1962), courtesy International Atomic Energy Agency.)

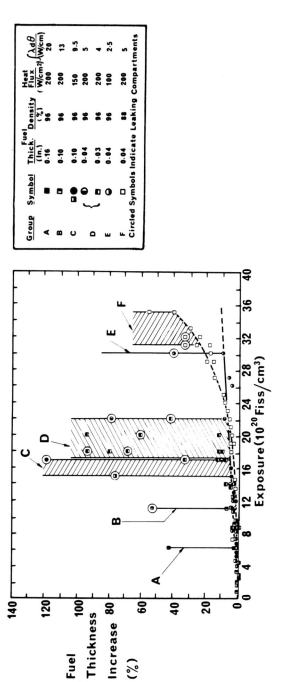

Group	Symbol	Fuel Thick. (in.)	Density (%)	Heat Flux (W/cm²)	$\left(\lambda\dfrac{d\theta}{dr}\right)$ (W/cm)
A	■	0·16	96	200	20
B	◨	0·10	96	200	13
C	◨ ●	0·10	96	150	9·5
D	{ ◨	0·04	96	200	5
	◑	0·03	96	200	4
E	◐	0·04	96	100	2·5
F	□	0·04	88	200	5

Circled Symbols Indicate Leaking Compartments

Fig. 3-16. Observed increases in plate thickness as a function of UO_2 exposure for diffe-rent combinations of plate thickness, fuel density, and surface heat flux; all compart-ments 1/4-in. wide. (After R. C. Daniel et al. (1962)).

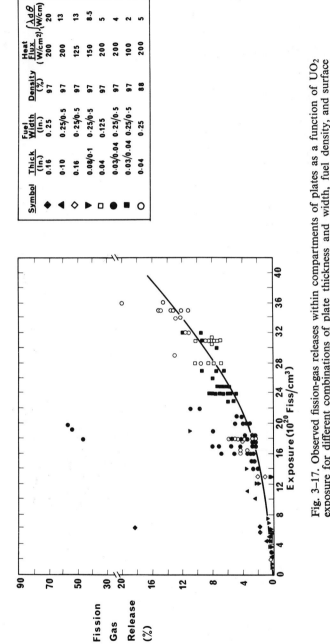

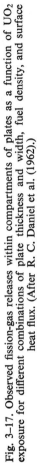

Fig. 3-17. Observed fission-gas releases within compartments of plates as a function of UO_2 exposure for different combinations of plate thickness and width, fuel density, and surface heat flux. (After R. C. Daniel et al. (1962).)

Symbol	Thick (in.)	Fuel Width (in.)	Density (%)	Heat Flux (W/cm²)	$\int \Delta d\theta$ (W/cm)
◆	0.16	0.25	97	200	20
◀	0.10	0.25/0.5	97	200	13
◇	0.16	0.25/0.5	97	125	13
▶	0.08/0.1	0.25/0.5	97	150	8.5
□	0.04	0.125	97	200	5
●	0.03/0.04	0.25/0.5	97	200	4
■	0.03/0.04	0.25/0.5	97	100	2
○	0.04	0.25	88	200	5

put and the fuel thickness. As discussed in Sec. 1.3, this provides a relative measure of the temperature drop in the fuel. There was little, if any, correlation between the gas release and the value of the integrated conductivity, probably because all the fuel temperatures were too low for temperature-sensitive release to be important (Sec. 3.6). Similarly, the first stage of the volume expansion showed no systematic dependence on the integrated conductivity. However, the integrated conductivity, and therefore the fuel temperature, had a large effect on the critical burnup, as can be seen by comparing groups of specimens in Fig. 3-16.

A similar comparison between groups may indicate that mechanical restraint on the fuel delayed breakaway. For other factors constant, the restraint imposed on the fuel is inversely proportional to the fourth power of the compartment width. Thus, later breakaway in compartments of 1/8-in. width than in those of 1/4-in. width suggested that restraint was effective, but there was no appreciable difference on going up in width to 1/2 in. Also, plates immersed in liquid metal at a pressure of a few atmospheres showed no more rapid bulging than those already described that were in pressurized water at more than a hundred atmospheres. Unfortunately, the reduced swelling of the compartments 1/8-in. wide may be due to lower fuel temperatures, since a larger fraction of the heat generated can be extracted through the side walls. Conversely, the UO_2 fragments in which gross swelling was attributed to high temperatures were unrestrained. Observations on cermets (Sec. 5.2) show that reducing the restraint, without appreciable increase in temperature, can accelerate breakaway swelling. The arguments made in connection with metallic fuels (Sec. 2.5) suggest that the level of external restraint normally expected from a sheath and coolant should not be significant during the early stages of swelling, until the volume increase has reached about 10%. With the sheath bulged away from the fuel, however, the sheath strength must be important.

Compartments with low density UO_2 (9.6 g/cm^3) did

not suffer breakaway as soon as those with UO_2 of normal density (10.6 g/cm^3) operating under similar conditions. However, uncertainties regarding temperature again confuse the situation. The low-density plates had been assembled by a pressure-bonding technique: Subsequent examination revealed that the sheath had been deformed into surface irregularities in the fuel, which gave a more intimate contact than for the higher density fuel. It is possible that consequent differences in the temperature drop across the fuel/sheath interface could account for the difference in behaviour. More likely, the pre-existing porosity accommodates some of the irradiation-induced swelling.

Thus, it appears that decreasing the fuel temperature, increasing the restraint on the fuel, or decreasing the fuel density may cause breakaway to be postponed to a higher burnup. However, it must be conceded that the independent effect of these factors acting separately has not been conclusively established.

Daniel et al. provided a possible interpretation of their results. During the first stage the very small volume increases can be explained simply by the greater atomic volumes of the fission fragments than of the uranium burnt. If the rare gases are precipitated, they must be at a very high pressure in extremely fine porosity. Even these small volume increases are not fully transmitted into sheath bulging until the original porosity in the UO_2 sinter has been filled. Meanwhile, a fraction of the fission-product gases has been released from the fuel, thereby generating a pressure within the sheath. When the pressure is sufficient to bulge the sheath away from the fuel, the temperature of the latter increases and, hence, more gas is released. Daniel et al. argued that this self-aggravating process represents breakaway and they developed an equation relating the critical burnup to other relevant factors.

Such a mechanism for breakaway would be specific to plates and might not apply to cylindrical rods, since the criterion for rapid swelling is sufficient internal pressure to separate fuel and sheath. There is, however, an unexplained inconsistency. One group of specimens began to fail by bulging on releasing one thirtieth of the *amount* of gas per unit volume of fuel found necessary to cause failure of another group with similar fuel density and restraint but with higher centre temperature. Also, serious swelling has been observed in the UO_2 particles of cermets without any separation of the fuel and sheath, so that the temperature remained below 1000°C (Sec. 5.2).

Therefore, it is possible that at high burnup a form of breakaway swelling occurs in the UO_2 itself, just as happens in metallic fuels. Pores are known to form under irradiation and to migrate in a thermal gradient: In UO_2 the thermal gradients are usually more severe than in metals. The rapid increase in compartment thickening would then represent the development of large pores, say over 1-μm diam. It is for this size of pore that external restraints begin to be effective in practice. Only *after* substantial swelling had already occurred would pores link up, releasing most of the gas from the fuel. Thereafter, sheath bulging might proceed catastrophically, as postulated by Daniel et al., because of a high internal gas pressure. Since this second mechanism would apply in modified form to cylindrical elements, it is of practical importance to establish which is appropriate.

Some rod elements achieved about 10×10^{20} fissions/cm^3 before failing (Gray and Mrazik, 1961), while others in the Yankee reactor have survived 27 000 MWd/tonne U (6.8×10^{20} fissions/cm^3) without failure (Naymark and Spalaris, 1964). In both lots the fuel was probably operating with higher centre temperatures than those of plate elements that had failed by swelling at the same burnup, but the sheaths were strong. The effective limit on burnup due to swelling in rod elements and its dependence on factors such as fuel temperature, external restraint, and fuel density have yet to be established.

It may seem paradoxical that swelling should occur at all in UO_2. Although there is no direct proof, it may be assumed that pressure of fission-product gases retained in the pores provides the driving force, as in uranium. Certainly there is no difficulty in explaining the origins of the pores, since they have been observed at a very early stage of development (Sec. 3.2). However, unlike uranium metal, UO_2 is able to release a large fraction of its gas at temperatures well below its melting point. At low temperatures, below 1250°C, most of the gas is retained in the UO_2 and Daniel et al. have shown that increasing the fuel temperature in this region encourages swelling. At temperatures over 1800°C most of the gas is released and swelling *may* be prevented. Thus, the propensity for swelling could pass through a maximum at some intermediate temperature. At least, there should be two competing effects; one releasing gas to a free surface, and the other causing swelling by the growth of the closed pores. If so, factors affecting the balance, such as the ratio of open-to-closed porosity, as well as thermal history and thermal gradients, could be vitally important to the onset of swelling. This uncertainty reinforces the need to understand the mechanisms of gas migration discussed in Sec. 3.6.

3.8 Migration and Segregation of Other Fission Products

The rare gases are by no means the only fission products that are mobile during irradiation. Cesium, iodine, and bromine are released on heating irradiated UO_2 and are among those detected in the coolant of elements with ruptured sheaths. Chemical analysis, autoradiography, microscopy, and electron microprobe analysis have all demonstrated segregation of fission products in the irradiated fuel.

Techniques were specially developed for procuring small samples of UO_2 from various positions across sections of irradiated elements (Bates et al., 1962b; Bates, 1964c). These samples were then analysed for plutonium and several fission products. Figure 3–18 illustrates the variations in the concentration of five isotopes. In this cross section, marked depletion of certain fission products has occurred from the hot annulus of UO_2 just inside the narrow columnar grains. In the porous central region that had probably been molten at the final shutdown, the concentration was appreciably higher. Other sections generally show depletion somewhere in the region of large grain growth, but sometimes exhibit more complex distributions. The concentration at the centre can be either a maximum or a minimum and, elsewhere, a minimum for one isotope often does not occur at the same radius as that for other isotopes. Some of the complications may be due to varying temperature distributions during the irradiation: If the central region melted at the start, but subsequently froze, a complex distribution of fission products might be expected. Leitz (1962) reported migration of cesium down a thermal gradient in the longitudinal direction of an element. The ^{134}Cs isotope was more mobile than the ^{137}Cs, as would be expected from the xenon precursor of the former being longer lived, if xenon is assumed to migrate more rapidly than cesium (but see results by Davies et al. below). Under the same conditions, strontium moved slightly, while cerium and zirconium did not.

Davies et al. (1963) measured the release of certain fission products from UO_2 during post-irradiation annealing in the range 1000 to 1600°C. They found the fractional releases of iodine, tellurium, and cesium were comparable with that of xenon, whereas those of strontium, barium, cerium, zirconium, and ruthenium were much lower, often by several orders of magnitude. These results may reflect differences in either diffusion rates or volatility: A given fission product can diffuse to a free surface but unless it (or an appropriate compound) is sufficiently volatile it will not be released. With high-density UO_2 (over 10 g/cm^3), the release of iodine, tellurium, and cesium was greater than that of xenon, especially at the higher temperatures. However, investigation of the effect over a wide range of exposures seems desirable, since the pores that delay the release of xenon (Sec. 3.6) may not affect the others to the same extent.

162

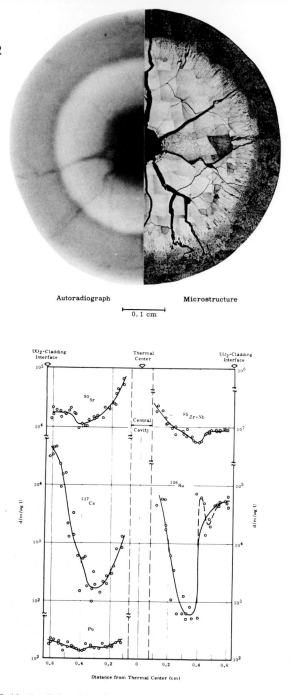

Autoradiograph   Microstructure

0.1 cm

Fig. 3–18. Radial analyses for certain fission products in an irradiated
UO$_2$ sinter whose autoradiograph and microstructure are shown
at the top: In the autoradiograph, black indicates activity. (Courtesy
J. L. Bates, Battelle-Northwest Laboratories.)

Autoradiography can provide dramatic confirmation of the segregation of fission products, but the isotopes primarily responsible for the film darkening have not been identified. In Fig. 3–18, the light band of low activity on the autoradiograph corresponds well with the minimum in the ^{106}Ru concentration: The ^{137}Cs minimum is also in the same region, but the concentration curve would erroneously predict higher activity in the outer region than in the centre. In other cross sections, the activity profile does not correspond to the concentration profile of any of the isotopes analysed.

Williamson and Hoffman (1963) obtained autoradiographs at different sections along a fuel element that had been irradiated with a high power per unit length. At the position of maximum power output, a very large central region was depleted of activity, but no region of equivalent enrichment was apparent on that section. At the position where the power output was 70% of the maximum, a narrow ring of strong enrichment immediately surrounded a central void. Uranium dioxide appeared to have sublimed from the hotter to the cooler end of the void, carrying with it a disproportionate amount of the volatile fissionproducts.

These observations serve as a reminder that fission products driven out of the hot region in one section need not deposit in the same section or even within the fuel. Cesium is sometimes found on the relatively cool sheath, and Fig. 3–18 indicates a gradient in the cesium concentration well out to the cool surface of the UO_2. Bain (1963) reported that the pattern of cracks in the UO_2 pellets was traced on the inner surface of the sheath by a deposit rich in ^{137}Cs. High concentrations of $^{140}Ba/^{140}La$ were found in dark ring deposits on the end faces of pellets. By passing irradiated Zircaloy-sheathed UO_2 elements in front of a collimated gamma counter, Lyons et al. (1964a) found high concentrations of ^{131}I and ^{132}Te at pellet interfaces, but only at the top of the stack. It is in this region that pellet separation sometimes occurs during irradiations with free-standing sheaths. Notably, the activity did not extend appreciably beyond the fuel into the end void.

Evidence for microsegregation is obtained from observations of white particles, generally a few microns in diameter, in polished cross sections of irradiated UO_2 (Fig. 3–19). These are commonly seen in the region of structural changes and, in some specimens, have been tentatively identified as beta uranium by x-ray diffraction (Roake, 1962; Bates, 1964b). The equivalent oxygen may have been removed by the sheath acting as a getter, or may merely have redistributed within the fuel under thermal gradients. These observations accord with earlier ones that heating unirradiated UO_2 under highly reducing conditions at temperatures in excess of 1800°C results in a precipitate of free uranium (MacEwan, 1962). For low exposures, the tarnishing and etching characteristics of the inclusions are very similar to those of bulk uranium, but by 10^{20} fissions/cm^3 the inclusions are extremely resistant to attack. Since such resistance has not been produced in bulk uranium by small amounts of alloying additions, there was a suspicion that fission products had become concentrated in the metallic-looking white phase. Also, autoradiographs showed spotty rings of high activity at the same radii as the rings of white particles. Confirmation has been provided by electron microprobe analysis. Roberts et al. (1964) found high concentrations of molybdenum and ruthenium in some of the particles and, less frequently, cerium and barium in others. Williams (1964) also found a high concentration of fission products in the inclusions, but he observed no comparable separation of the species. Since some sections have exhibited more than one distinct ring of inclusions, it seems that certain conditions of thermal gradient, or thermal history can effect a separation. Inclusions of similar appearance, but due to iron and nitrogen impurities introduced during fabrication, are sometimes observed. These may confuse the interpretation but are not pertinent to fission-product migration.

Thus, in UO_2 under irradiation at temperatures high enough to produce structural changes, gross migration of the fission products occurs, but, at lower temperatures,

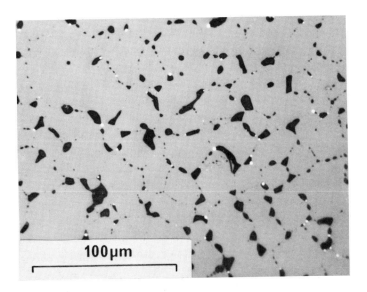

Fig. 3–19. Micrograph showing white particles in irradiated UO_2 that has been hot enough to undergo structural changes. (Courtesy A. S. Bain, Chalk River Nuclear Laboratories.).

their location is not properly established. Some segregation at grain boundaries is suggested by the enhanced etching of the boundaries in material irradiated to roughly 10^{21} fissions/cm^3 and of the subgrain boundaries in material at higher burnup (Fig. 3–5). Padden and Schnizler (1962) found the etching solutions to be enriched in ^{137}Cs for both conditions. Since there was no trace of the original boundaries in the subdivided structure, fission products had presumably migrated from them into the newly developed subgrain boundaries. At the other extreme of the temperature range, segregation at the subgrain boundaries of regions believed to have been molten (Fig. 3–3) has been demonstrated by high-resolution autoradiography (Bain et al., 1964b).

Berman (1963) performed experiments to investigate the segregation in material that had been irradiated at low temperatures. He pulverized irradiated sintered UO_2, then dissolved it progressively over two days and analyzed the

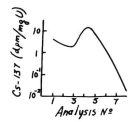

solution every six hours. Initially, the solution contained fission products in about the same concentration, relative to the uranium, as in the complete specimen. The concentration of fission products rose to a very high maximum at the fourth analysis and then decreased to a final value well below the average. Believing that the pulverizing stage had reduced the sinter to its component grains, since irradiated UO_2 fractures predominantly along grain boundaries, Berman interpreted the results as confirming segregation at the grain boundaries. He supposed that relatively rapid dissolution of fines accounted for the first few analyses. However, the specimen that gave the most consistent results had been irradiated for only a few hours at a temperature almost certainly under 650°C. The implied migration of a large fraction of certain fission products over distances of several microns is surprising. Other specimens gave confused results that were not explained. These experiments indicate that segregation occurs, but the location of the fission products deserves further investigation.

Shiriaeva and Tolmachev (1957) annealed irradiated UO_2 powder (of unspecified characteristics) and subsequently leached the surface of the particles without dissolving them. After only a few minutes in vacuum at 1200°C, most of the ^{99}Mo had apparently segregated at the surface, as would be expected of an insoluble species able to migrate. In porous specimens, therefore, solid fission products might be expected to migrate to the surfaces of pores: In fact, the white particles are often found associated with pores (Fig. 3–19).

Migration of the plutonium is considered in Sec. 3.12, as relevant to the behaviour of UO_2–PuO_2.

3.9 Crystal Structure

Irradiation damage to the crystal structure of UO_2 has been most thoroughly investigated by Fox et al. (1963). The observed increases in cell size are plotted as a function of exposure in Fig. 3–20. Increasing the irradiation temperature from somewhere in the range 65 to 100°C to 400°C

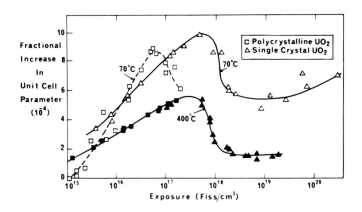

Fig. 3–20. Change in unit cell parameter of UO₂ on irradiation at 70 and 400°C. (After Roberts et al. (1964), courtesy United Nations.)

decreases the magnitude of the change but does not alter the shape of the curve. At the lower temperatures, the maximum change in polycrystalline specimens occurred at an exposure one tenth that for the maximum in single crystals: At 400°C, there was no detectable difference between the two materials, but none of the polycrystalline specimens had been irradiated above 2×10^{17} fissions/cm³. It is tempting to attribute the difference to grain-boundary effects, but the possibility of other factors being responsible cannot be excluded, since the two materials were prepared by completely different routes. Fox et al. also measured line broadening for single crystals irradiated at the lower temperatures. The line width gradually increased to four times its unirradiated value, then rapidly decreased by a factor of 2 in the range 1.7 to 9.1×10^{18} fissions/cm³, i.e., at an exposure significantly above that at which the cell dilatation dropped.

Such other observations as have been reported are in general agreement. Bloch (1961) observed a fractional increase of 8.6×10^{-4} in the cell size of polycrystalline UO₂ that had been exposed to 1.5×10^{17} fissions/cm³ at some temperature between 20 and 62°C. In another laboratory, the cell size of UO₂ dispersed in Al₂O₃ was found to increase with increasing exposure at temperatures under 100°C, to a dilatation about 10^{-3} at 2×10^{17} fissions/cm³ in the UO_2 phase (Berman et al., 1960). For samples of polycrystalline UO_2 irradiated at temperatures over 400°C to exposures from 0.6 to 3.6 $\times$ 10^{21} fissions/cm³, any change in cell size was less than the experimental uncertainty of 1 in 10^3 (Bleiberg et al., 1962). Bates et al. (1962) measured a fractional increase in cell size of 2.4×10^{-4} in a single crystal of UO_2 that had been irradiated to 1.44×10^{20} fissions/cm³ at temperatures not exceeding 400°C.

Post-irradiation annealing of the damage was also studied by Fox et al.. Both single crystals and polycrystalline material irradiated at temperatures under 100°C exhibited partial recovery in the range 200 to 400°C. Below 2×10^{16} fissions/cm³, two separate stages were apparent, near 250 and 400°C, but, for higher exposures, recovery was gradual over the whole range. In the most highly irradiated sample, a single crystal exposed to 2×10^{18} fissions/cm³, a reflection profile was observed to broaden and become asymmetric even below 200°C, where recovery first became appreciable. The asymmetry suggests that recovery does not proceed uniformly throughout the sample, a possibility that would also explain the failure to resolve two separate stages in this range. All five specimens annealed had a further recovery stage between 550 and 700°C. By 950°C, there was no residual dilatation in the single crystals, but approximately one third of the original damage remained in the polycrystalline specimens. Bloch's sample, exposed to 1.5×10^{17} fissions/cm³, showed closely similar annealing behaviour up to 700°C, but at higher temperatures recovery continued in her sample, while the remaining damage was unchanged in polycrystalline samples of Fox et al. up to 900°C. Bates (1964a) reported that a polycrystalline sample that had been irradiated to 1.4×10^{19} fissions/cm³ below 100°C suffered a fractional decrease in cell size of 6.5×10^{-4} during annealing in vacuum for eight hours at 960°C, thereby regaining its unirradiated value.

It is generally supposed that the initial increase in cell

size at low exposures is due to point defects, or possibly a few associated defects, straining the lattice. Fox et al. attributed the maximum in the dilatation versus exposure curve to growth of defect clusters, with a consequent reduction in the concentration of point defects. The lattice strain would qualitatively account for Ross's observation of increasing damage to the thermal conductivity of polycrystalline UO_2 with increasing exposure (Sec. 3.4). However, he found saturation in the damage rather than a maximum, and the saturation occurred at an exposure an order of magnitude lower than that for the maximum in the dilatation. Fox et al. supposed that some growth or ordering of the clusters followed the maximum in the dilatation, but electron microscopy reveals no change in the structure until subdivision of the original grains occurs about 3×10^{21} fissions/cm^3 (Sec. 3.2). If the measurements of different properties could be made on samples of the same material, all irradiated at the same time, direct comparison might permit identification of the various defects, as has been attempted with UC (Sec. 4.3).

Electron microscopy (Sec. 3.2) and the trapping of fission-product gases (Sec. 3.6) indicate that a high concentration of gas-filled pores exists after annealing irradiated UO_2 at 1400°C. The x-ray studies show that full recovery of the damage to the crystal structure can have occurred at lower temperatures. Thus, the various recovery steps observed by x-ray techniques presumably represent stages in the annealing of dislocation loops and the growth of pores from point defects, but the separate mechanisms have not been identified. At the lower temperatures, point defects would be expected to be mobile, while the final stages may be due to migration of very small pores by surface diffusion of the atoms immediately surrounding them (Sec. 3.6). Solution of some fission products in the UO_2 lattice would cause a decrease in cell size, while others would cause an expansion. Although the net effect cannot be calculated, a change in cell size might be expected to accompany the precipitation of solid fission products (Sec. 3.8).

3..10 Other Properties

Hardness

One of the properties of UO_2 affected by irradiation is its hardness. However, before attributing any apparent change to irradiation, it is necessary to make adequate allowance for other factors known to affect the hardness value, such as O/U atomic ratio, bulk density, orientation of the crystal face tested, rotation of the indenter, and the applied load.

After irradiating single crystals at temperatures under 100°C, Bates (1963) observed an increase in their hardness measured at room temperature, possibly saturating with increasing exposure, in the range 10^{14} to 6×10^{15} fissions/cm^3 (Fig. 3–21). Another of his single crystals exhibited an increase of 18 to 20% after 1.4×10^{20} fissions/cm^3, but it cannot be grouped with the others, since its estimated irradiation temperature was between 325 and 400°C. Bates performed a survey across a section of a sintered pellet that had been exposed to 2×10^{19} fissions/cm^3 and found that the hardness had increased by about one third over most of the cross section. However, the increase was only half as much in the innermost part of the central region that had been molten during irradiation. (This interpretation of the microstructure is the one given in Secs. 3.2 and 3.3 but differs from that of Bates, who believed that the melting point had not been exceeded.) Thus, at this level of exposure, the increase in hardness is not sensitive to irradiation temperature from approximately 500°C to the melting point, but observations on other properties (Secs. 3.4 and 3.9) discourage the assumption that such insensitivity will necessarily extend to lower temperatures. An increase of one third in the hardness of a sinter exposed to 2×10^{20} fissions/cm^3 at 450°C (Padden, 1961) suggests that polycrystalline material may be susceptible to greater irradiation hardening than single crystals. Another study (Padden and Schnizler, 1962) of sintered UO_2, irradiated between 400 and 1200°C, showed a decrease in the hardness between 1.3×10^{21} and 3×10^{21} fissions/cm^3, so that the hardness

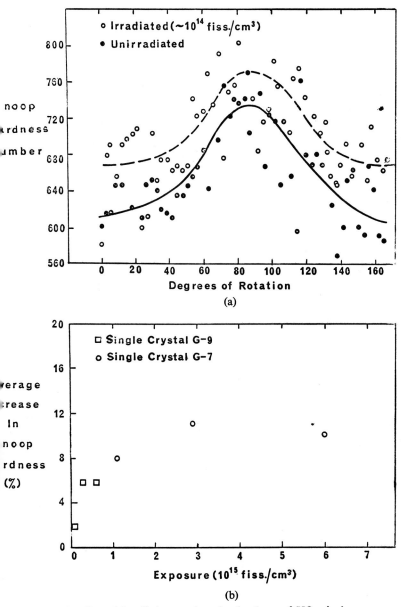

(a)

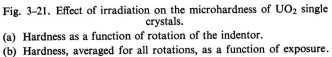

(b)

Fig. 3–21. Effect of irradiation on the microhardness of UO₂ single crystals.
(a) Hardness as a function of rotation of the indentor.
(b) Hardness, averaged for all rotations, as a function of exposure.
(After Bates (1963).)

reached roughly the same value as that of unirradiated material by the highest burnup. The authors associated the decrease with the development of a fine substructure in the irradiated material (Secs. 3.2 and 3.8).

The causes of the hardening have not been identified. Presumably at low exposures and at low temperatures the effects of point defects and dislocation loops predominate. At higher exposures, the fission products present as individual atoms may be expected to contribute; at higher temperatures, precipitated fission products, both solid and gaseous, could be effective.

Stored Energy

That part of the fission energy causing defects and disorder in the crystal lattice is not immediately available as heat but is retained as stored energy in the fuel. During annealing of the damage, the stored energy is converted to heat. In practice, only some of the annealing stages liberate enough heat for detection even by sensitive calorimetry.

Childs (1962) studied the release of stored energy from various uranium oxides that had been irradiated at temperatures under $55°C$ to exposures up to 10^{18} fissions/cm^3. In single-phase UO_2, no detectable release was observed up to $750°C$, the maximum annealing temperature in these experiments: The estimated limit of detection was 0.5 cal/g, if released in a temperature interval of 100 deg C. Thus, annealing stages below $750°C$ in the damage to lattice parameter (Sec. 3.9) and thermal conductivity (Sec. 3.4) are associated with only small energy releases. U_4O_9 exposed to 10^{18} fissions/cm^3 gave a small release, about 1.5 cal/g, at $670°C$ and possibly an even smaller one at $150°C$.

Specimens of UO_2 containing U_4O_9 precipitate exposed to the same level produced a release of 2.4 cal/g at $340°C$, where neither of the constituents had shown any effect. Childs suggested that irradiation had converted the whole sample to the UO_2 structure stable at higher temperatures, possibly by partial homogenization of the oxygen, and that the release represented reprecipitation of the U_4O_9.

A similar small release at 520°C from UO_2 containing precipitated uranium metal was tentatively attributed to the reprecipitation of the uranium that had been dissolved during irradiation.

Much larger releases were obtained from irradiated U_3O_8. As exposure was raised from 10^{14} to 10^{16} fissions/cm^3 the energy release increased from 2 to 25.5 cal/g, while the temperature of release increased from 150 to 350°C. At higher exposures, up to 10^{18} fissions/cm^3, the temperature of the major release and the total heat released were unchanged, but a small part of the heat was released between 100 and 350°C. X-ray diffraction, performed on the most highly irradiated specimen, showed no trace of the original lines, but these were found to have been restored after subsequent heating to 720°C. Boyko et al. (1958) observed diffraction lines of U_3O_8 still present, but considerably reduced in intensity, after an exposure of 2×10^{18} fissions/cm^3 at temperatures around 100°C.

Childs interpreted his results in terms of a model in which each fission fragment damaged a certain volume of the lattice. The eventual absence of diffraction lines and the similarity in magnitude between the stored energy and the latent heat of fusion suggested that the damaged region had a structure close to that of the liquid. Analysis of the results yielded a value for the damaged volume of 2.7×10^{-16} cm^3/fission fragment. Childs supposed that at low exposures, where regions of the original U_3O_8 structure still existed, oxygen self-diffusion would suffice to restore the transformed material, but at high exposures prior nucleation of the U_3O_8 structure would be required. The increase in annealing temperature would be a consequence of such a mechanism. If the irradiation were to be performed at a temperature at which appreciable annealing occurs, the disordering might be expected to require a greater exposure or to be absent altogether. A similar disordering of the structure during irradiation at low temperatures possibly occurs in U_3Si (Sec. 4.7), but the stored energy in that material has not been measured.

Electrical Properties

Amelinckx and his colleagues (1965) measured the electrical conductivity of UO_2 single crystals that had been irradiated at 85°C. In a specimen whose composition was very close to stoichiometric, so that no U_4O_9 precipitates were detectable even by electron microscopy, the conductivity was unchanged up to an exposure of 1.6×10^{16} fissions/cm^3. However, the authors were concerned that in stoichiometric crystals, where the apparent conductivity is very low, conduction through a thin surface layer of higher oxide may limit the true bulk conductivity that can be measured. Where the oxygen/uranium atomic ratio exceeded 2.001, irradiation decreased the conductivity at room temperature by about three orders of magnitude. The damage appeared to have saturated by 10^{16} fissions/cm^3 suggesting that all charge carriers originally present along a fission-fragment track had been trapped. Such trapping also explained the increase in the temperature coefficient of the conductivity observed following irradiation. Annealing of the damage occurred at temperatures above 700°C, with an activation energy of 2.5 eV. Since the activation energy for self-diffusion of uranium (the sum of terms for formation and migration of vacancies) is around 4 eV, uranium vacancies acting as traps for electron holes are probably responsible for the damage to the electrical conductivity.

Most nominally stoichiometric specimens have an oxygen/uranium atomic ratio greater than 2.001. Some sinters irradiated by Roake (1962) at temperatures under 100°C had exhibited decreases in conductivity for exposures up to 4×10^{18} fissions/cm^3, in a manner similar to that for the hyperstoichiometric specimens of Amelinckx et al. After 1.56×10^{19} fissions/cm^3, however, the conductivity was greater than that of the unirradiated material for all temperatures up to 800°C, and, as a result of annealing during the first measurement run, the conductivity was even greater in a second run. This increase might be due to irradiation-induced solution of the U_4O_9 precipitate, but some results by Amelinckx et al. indicate that the U_4O_9 is reprecipitated at 500°C during post-irradiation annealing. Roake noted that the electrical measurements showed none of the annealing steps found for the thermal conductivity of the same specimens (Sec. 3.4), confirming the small electronic contribution to the thermal conductivity.

Karkhanavala and Carroll (1961) measured the electrical conductivity of sintered UO_2 during irradiation, in the range 200 to 750°C. They reported an increase in the conductivity, by a factor of about 5, over the whole range after an exposure of 10^{17} fissions/cm^3, but, since their values included the contact resistances, the conclusion is open to question. Amelinckx et al., using a method that eliminated the contact resistances, observed conductivity decreases similar to those in post-irradiation measurements for a slightly hyperstoichiometric single crystal during irradiation at 88°C. The absence of any detectable increase in the conductivity at the start of the irradiation argued that charge carriers due to ionization must have a very short lifetime.

Measurements at low temperature showed very slight increase in the susceptibility of UO_2 and a broadening of the maximum at the Néel point as a result of irradiation (Weil and Cohen, 1963). Since oxygen interstitials have the opposite effect, the authors concluded that some other defect must be responsible.

3.11 Powder Compacts

Although most of the experience of irradiating UO_2 has been obtained with sintered pellets, many elements containing powdered UO_2 have also been tested. The feed material, though normally termed "powder", is often of graded-particle-size distribution to facilitate compaction, with the particles themselves being of near theoretical density. After the fuel has been loaded in the sheath, the latter is vibrated to effect consolidation: In some production routes, the elements are then swaged to achieve a higher fuel density, with associated reduction in sheath diameter. Fuel elements containing compacted UO_2 powder usually have fuel densities from 85 to 95% of that for fully dense UO_2.

Unfortunately, many of the irradiations of UO_2 powder compacts have been intended as proof tests of specific designs. The results may be important to the designers concerned, but little information of general application has been gained. Also, interpretation from post-irradiation examination is more difficult for powder compacts because of the greater changes that occur during irradiation. From observations of cracks in irradiated compacts, Robertson et al. (1958) concluded that under irradiation the individual particles sintered together at temperatures as low as 800°C, compared with temperatures around 1500°C required outside the reactor. Figure 3–22 is a micrograph illustrating sintering of adjacent particles in the outer cool region of a fuel element.

The enhancement of sintering under irradiation is presumably a consequence of the large deposition of energy by fission fragments in the surface layers of the particles: Contact points become welded together while any ejection of UO_2 from the surface (Sec. 3.6) is equivalent to rapid surface diffusion. Laboratory measurements show the powder compact to have lower thermal conductivity than a sinter of the same density. Thus, sintering, wherever it occurs, should increase the conductivity. As the centre temperature is increased in UO_2 powder compacts, equiaxed grain growth, columnar grain growth, and melting can all occur, yielding structures similar to those seen in sintered UO_2 (Sec. 3.2). Once these structural changes have taken place, the thermal conductivity of the UO_2 is probably independent of whether the fuel was initially a sinter or a powder compact.

Bain (1961) confirmed the low conductivity under irradiation from measurements of the extent of central melting in powder compacts that were irradiated for only a few minutes. However, other powder specimens that had been fabricated with central thermocouples demonstrated an increase in conductivity over an exposure of several hours, even with centre temperatures as low as 300°C

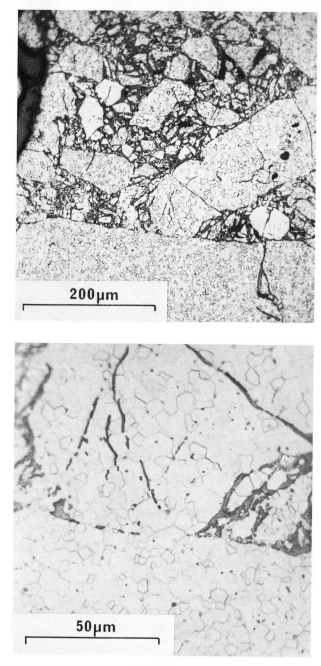

(a)

200µm

(b)

50µm

Fig. 3–22.
(a) Micrograph of UO₂ powder compact that was irradiated at an estimated temperature about 400°C.
(b) Micrograph at higher magnification showing in-reactor sintering of adjacent particles.

(After Flaherty (1963).)

(Roake, 1962). When the effects of irradiation on the conductivity of sintered UO_2 (Sec. 3.4) are also considered, it is apparent that the conductivity of UO_2 powder compacts must be a complex function of the material's thermal and irradiation history as well as of the temperature of measurement.

The accumulated release of fission-product gases from powder compacts has been measured for many individual elements. In general, the release increases with increasing power output per unit length of fuel and, thus, with increasing fuel temperatures. The release may also be expected to depend on the initial density and on how much in-reactor sintering has occurred before substantial amounts of fission-product gas are generated. Several in-reactor failures of elements containing compacted UO_2 have been attributed to excessive internal gas pressure, but the gas responsible was a naturally occurring one and not a fission product. The high specific surface area of the powder can permit adsorption during fabrication of a large volume of gas that is driven off when the fuel heats up under irradiation. Also, early batches of fused UO_2 sometimes contained gaseous impurities that were evolved on heating.

In comparing the thermal expansion of powder compacts with that of elements containing sintered pellets, the basis of comparison should be clearly stated. If two elements have the same power output per unit length, the compact would have the higher fuel temperatures because of its lower conductivity. Usually the powder compact has a lower density than the sinter and may accommodate some of its own expansion by sintering. However, the sintered pellets may have dishing on their end faces and a diametral clearance between fuel and sheath, which are absent in the powder elements. Pashos et al. (1964) summarized their experience as showing greater diametral expansion for powder compacts than for pellets rods, when the power output per unit length was the same for both, but Feraday and his colleagues (1964, 1965) reported the reverse. Differences in the rate of heating the fuel probably do not explain the contradiction, since some of Feraday's specimens attained their operating temperatures very rapidly, and this is the condition likely to cause most expansion before fuel densification can occur. Details are lacking to determine whether differences in fuel density or in the extent of central melting could reconcile the results.

Initially, there was concern that much of the UO_2 powder would escape into the coolant, thereby contaminating the system if the sheath ruptured during irradiation. Again, the rapid in-reactor sintering reduces any original differences between powder compacts and presintered pellets. Thus, several powder elements have operated with deliberate or accidental holes in their sheaths without releasing a large amount of fuel. It is true that major releases of what was originally UO_2 powder have occurred when the sheath split wide open on rupturing, but fragments of sintered pellets behave similarly.

The resistance of UO_2 to corrosion by most potential coolants played a large part in its original selection as a fuel. After prolonged exposure to water under irradiation, oxidation of the UO_2 has been observed (Robertson et al., 1958; Harder and Sowden, 1960), but the effect does no appear to be markedly enhanced in powder compacts. In fuel elements that have operated with water in contact with the fuel, a U_4O_9 precipitate is usually present in the UO_2: Yellow specks, possibly $UO_3 \cdot 2H_2O$, are sometimes seen. Since the oxygen comes from radiolysis of the water, sheathing materials that absorb hydrogen, e.g., zirconium alloys, probably favour the oxidation.

3.12 Mixed Oxides

There have been three main reasons for interest in mixed oxides as potential fuels:

1. To improve some specific property of UO_2, such as thermal conductivity or capacity for gas retention. Usually only minor additions, of the order of one molecular percent of another oxide, are considered for this purpose.

2. To combine the most suitable fissionable and fertile materials in the optimum ratio. For example, around

20% PuO_2 in UO_2 is a possible fuel for fast reactors, while around 2% UO_2 (fully enriched in [233]U or [235]U) in ThO_2 may be attractive in certain thermal reactors.

3. To dilute highly enriched fissionable material. In some applications the fissionable material should not be associated with any fertile material. It is therefore diluted in another oxide, so that the heat fluxes do not exceed the heat extraction capabilities of the coolant. The fissionable oxide, UO_2 or PuO_2, is usually the minor component, and the diluent an oxide of some metal with a low-capture cross section for neutrons.

The low thermal conductivity of UO_2 has naturally inspired attempts at improvement. Tennery (1959) suggested that additions of oxides of tri- or penta-valent metals might contribute enough free electrons to affect the heat transfer significantly by that means. Following some encouraging laboratory measurements, samples of hydrogen-sintered UO_2–4mol% Y_2O_3 were compared with UO_2 controls in an irradiation experiment. Similar specimens of each were exposed to the same power per unit length for brief periods and the extent of central melting produced was subsequently measured. The comparison showed the Y_2O_3 to have no beneficial effect on the thermal conductivity and further laboratory tests confirmed the result (Robertson et al., 1960; Powers et al., 1960). The discovery of enhanced conductivity in hypostoichiometric UO_2 (Sec. 3.4) stimulated May and Stoute (1965) to reexamine the effect of additives. Using their simple comparator, they found that partially reduced UO_2–0.1mol% Y_2O_3 and UO_2–0.1mol% La_2O_3 exhibited increased conductivity similar to that of partially reduced UO_2, but that the enhancement became apparent at lower temperatures. Measurements under irradiation now seem desirable.

Extensive open porosity in UO_2 sinters probably facilitates gas release. Since small amounts of TiO_2 and Nb_2O_5 can act as sintering aids, the effect of these additives on gas release was tested in a comparative irradiation. MacDonald (1963) found that specimens of UO_2–0.35 mol% TiO_2 and UO_2–0.4mol% Nb_2O_5 released just as much gas as control specimens of UO_2 when irradiated under similar conditions. However, he did observe an interesting migration of niobium to the hot centre, where it apparently formed a eutectic in the UO_2–Nb_2O_5 system.

ThO_2–UO_2

ThO_2 and UO_2 are isomorphous with little difference in lattice parameter, so a single phase can exist over the entire composition range. However, fabrication from the individual powders does not always achieve equilibrium, and segregated particles of ThO_2 and UO_2 are sometimes observed in sinters. With ThO_2-rich compositions air sintering is possible, but yields a product with greater oxygen content than does hydrogen sintering.

Irradiation experience with ThO_2–UO_2 fuels has been mostly with compositions of 10wt% UO_2 or less, but some specimens containing up to 50wt% have been tested. The irradiations have demonstrated behaviour qualitatively similar to that for pure UO_2, but few direct comparisons have been possible. From measurements of the extent of central melting observed in irradiated specimens of ThO_2–25wt%UO_2 prepared by air sintering mixed ThO_2 and U_3O_8 powders, Neimark (1962) deduced that the mean thermal conductivity from about 300°C to the melting point was appreciably lower than that of material obtained by hot pressing mixed ThO_2 and UO_2 powders: He attributed the effect to the higher oxygen content of the former. Neimark (1962), Rao (1963), and Neimark and Kittel (1964) all noted a characteristic metallographic structure of the formerly molten zone, whose appearance differed from that of the equivalent zone in pure UO_2 (Fig. 3–3 in Sec. 3.2). Columnar grains, sometimes associated with lenticular pores, have been observed in ThO_2–UO_2 as in UO_2 (Sec. 3.2): Neimark suggested that the lower vapour pressure of ThO_2 at any temperature should cause the columnar grains to form at a temperature about 350°C higher than in UO_2. Equiaxed grain growth and thermal stress cracking are similar to the corresponding effects in UO_2.

Rao (1963), who reviewed the earlier experience, performed a series of irradiations in which he measured the extent of melting in various compositions up to 19wt%UO_2. The exposures were of short duration (3 min.), but the results permitted comparison of the integrated thermal conductivity with respect to temperature from about 400°C to the melting point under closely controlled conditions. He found that the conductivity decreased as the UO_2 content increased from 5.4 to 19 wt% in hydrogen-sintered specimens. Compositions in the range 5 to 10wt% UO_2 possessed conductivities similar to that of pure UO_2, while extrapolation indicated that pure ThO_2 would have a value at least ten percent higher. When allowance was made for the differences in thermal conductivity, no difference in expansion behaviour was observed between the various compositions, including pure UO_2.

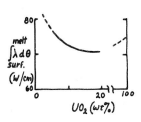

The potential of ThO_2–UO_2 as a 'nuclear fuel was demonstrated by Neimark and Kittel (1964), who irradiated, in cold water, pellets lead bonded to aluminum-alloy sheaths. They prepared sinters of three compositions, 6.36, 12.7, and 25.6 wt% UO_2, all of low density (69 to 82% of theoretical), from mixtures of ThO_2 and U_3O_8: The oxygen-to-metal atomic ratio was therefore over 2, and probably about 2.05. A burnup of 33 200 MWd/tonne oxide (8.5×10^{20} fissions/cm^3) was achieved in ThO_2–25.6wt%UO_2 with a power output of approximately 550 W/cm of fuel. Other specimens operated at power outputs as high as 750 W/cm but during the final reactor cycle only. Below 400 W/cm, the increase in sheath diameter was only about one third of one percent, but at 550 W/cm and above it was a few percent. There was no evidence for the diametral expansions increasing with increasing burnup as would be expected for fuel swelling. The release of fission-product gases increased with power output, reaching 48% at 550 W/cm and even 81% for one of the specimens overheated in the final cycle. The authors attributed the large releases to the excess oxygen and the low density of the fuel. Both the expansions and the gas releases were significantly greater than would have been obtained with high-density stoichiometric UO_2, but a substantial improvement in performance should be possible by using high-density stoichiometric ThO_2–UO_2 of lower UO_2 content.

The successful irradiation of sintered ThO_2–4.45wt%UO_2 to a burnup of 95 000 MWd/tonne (U + Th) (2.4×10^{21} fissions/cm^3) at a power output of approximately 500 W/cm of fuel was reported by Rabin and Ullmann (1964). The fission-product gas release was 12.4% of that generated. Initial results showed that any swelling did not exceed 0.5% volume increase per 10^{20} fissions/cm^3, in a sheath whose thickness was one tenth of the fuel diameter. Grain boundaries were still apparent in the fuel of a specimen irradiated 1.3×10^{21} fissions/cm^3, but they had disappeared by 2.4×10^{21} fissions/cm^3 although the material still gave an x-ray diffraction pattern (cf, ZrO_2–UO_2). At the lower exposure no change in the lattice parameter was observed, but at the higher exposure a decrease of about one part in a thousand, close to the experimental error for the equipment used, was reported.

Sintered ThO_2–10wt%UO_2 was irradiated in deliberately punctured sheaths in the Experimental Boiling Water Reactor at relatively low power outputs (up to 150 W/cm of fuel). R. F. S. Robertson (1960) reported that the rare gases and their decay products constituted the major activity release: The release of radioiodines was notably much less than from UO_2 fuel elements irradiated with punctured sheaths in pressurized water. Although it is tempting to attribute their absence to differences in properties of the fuels, the boiling environment or a diminished incidence of power cycling may have been responsible.

To improve the thermal conductivity of hot-pressed pellets of ThO_2 containing 10, 30, or 50 wt% UO_2, Neimark et al. (1961) incorporated randomly oriented fibres of molybdenum or niobium. Post-irradiation examination showed that the niobium had reacted with the fuel. The presence of molybdenum fibres had substantially suppressed thermal cracking and other structural changes, but no direct comparisons of thermal conductivity were possible. In the region of columnar grains the molybdenum had migrated to the centre and agglomerated by some mechanism not fully understood.

$UO_2–PuO_2$

Continuous solid solubility exists between the two components in the pseudo-binary system $UO_2–PuO_2$, but, under conditions of oxygen deficiency, separation into two phases, UO_2 and Pu_2O_3, can occur. In practice, three levels of plutonium content are of interest: around 20% for fast reactors; a few percent as enrichment in thermal reactors; and under 1% produced in uranium during irradiation.

Sintered specimens containing deliberate additions of 1% plutonium were irradiated to nearly 10^{20} fissions/cm³ (approximately 3000 MWd/tonne fuel) by Sayers and Worth (1961). These had a power output up to 400 W/cm length of fuel and a fuel-surface temperature estimated to be around 1000°C. The percentage release of gaseous fission products, as well as the type and extent of grain growth, did not differ significantly from those in control specimens of pure UO_2. Similarly, Frost et al. (1962b) compared sintered specimens containing ten percent plutonium with pure UO_2. Again, there was no noticeable effect of the plutonium, but these specimens operated at a lower power output (under 180 W/cm of fuel) and with a lower surface temperature, so that no grain growth was observed and gas release was no more than a few percent. Although exposures were up to 4.5×10^{20} fissions/cm³ (approximately 15 000 MWd/tonne fuel), no evidence of swelling associated with gas-filled porosity was seen.

Gerhart et al. (1961) irradiated both sintered and swaged $UO_2–20wt\%PuO_2$ to very much higher exposures—up to 125 000 MWd/tonne fuel. The sheath temperature was around 550°C and the power output, up to 500 W/cm of fuel, was sufficient to cause extensive structural changes in the fuel and large gas releases. However, no controls of pure UO_2 were included and relatively large variations in fuel density, fuel/sheath diametral clearance, sheath temperature, and duration of irradiation prevented quantitative comparison with the behaviour of UO_2. The length of the fuel, which was constrained radially by a strong sheath, increased by as much as 13% of the original length as a result of the irradiation, but inability to correlate the changes with fuel burnup or power output prevented any useful conclusions regarding fuel swelling. Thus, while this series of tests demonstrated the feasibility of operating short lengths of $(U, Pu)O_2$ to high exposures in a strong sheath, it also pointed out the need for more quantitative measurements on the material's behaviour.

The release of gaseous fission products from $(U, Pu)O_2$ sinters during irradiation as a function of the power output per unit length has been studied under more closely controlled conditions by Bailey and Chikalla (1964). For 90% dense pellets the shape of the resulting curve was indistinguishable from that obtained with pure UO_2 (Sec. 3.6): Up to a critical power per unit length the release was small, but thereafter it increased approximately linearly (Fig. 3–23). Pellets whose original density was only 60% of that theoretically possible gave greater releases and more scatter on the same plot. The migration of solid fission products was investigated by analyzing cores drilled at intervals along a diameter on a polished cross section. Concentrations of ^{137}Cs were generally depleted in regions of structural change and correspondingly enhanced further out: ^{106}Ru behaved similarly, except that its concentration passed through a maximum somewhere in the columnar grains: ^{90}Sr sometimes showed slight evidence of movement, but $^{95}Zr/^{95}Nb$ and $^{144}Ce/^{144}Pr$ remained *in situ*. Experience with UO_2 (Sec. 3.8) warns that considerable variations in the distribution may occur between specimens having similar structures but different histories.

Although there is evidence that plutonium can migrate

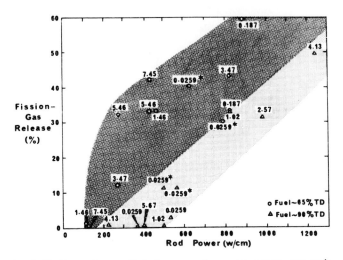

Fig. 3–23. Dependence of fission-gas release on power per unit length of UO_2–PuO_2 sinters for two different densities. The number alongside each point is its PuO_2 content in molecular percent: An asterisk indicates coprecipitated $(U, Pu)O_2$, all others mixed oxides. (After Bailey & Chikalla (1964).)

within the UO_2, the necessary conditions have not been established, and sometimes no detectable segregation is observed. Morgan et al. (1961) analyzed fragments of UO_2 pellets for the plutonium produced during irradiation. In one specimen, samples taken from just outside the region of obvious grain growth contained less plutonium than those taken from further out or further in: The effect was reproducible on other sections within the same specimen, but was not observed in other specimens. Using a more elegant technique for sampling, Bates et al. (1962 b) found a minimum in the plutonium concentration for an irradiated specimen of vibratory compacted UO_2, which occurred in the region of columnar grains and outside the zone that had been molten (Sec. 3.2). In a specimen of swaged UO_2 in which columnar grains extended to the centre, they observed a central maximum in the plutonium concentration. Figure 3–18 (Sec. 3.8) is reasonably typical of these specimens in showing a slight, but probably significant, migration of the plutonium. In other specimens, no segregation has been detected. Leitz (1962) reported no significant variations in the plutonium concentration (within 20% experimental error) in those specimens of Gerhart et al. that were exposed to the highest burnup, even though they exhibited extensive structural changes.

Runnalls (1962) described the irradiation of sinters of the mixed oxides containing two percent plutonium: In some specimens the plutonium was deliberately segregated in coarse refractory particles of PuO_2 during fabrication, while it was homogeneously distributed through the UO_2 in others. For nearly the same power output per unit length (675 W/cm of fuel), a heterogeneous specimen showed less grain growth and released only half as much fission-product gas as a homogeneous one. The comparison suggested that the solution of PuO_2 in UO_2 may significantly reduce the latter's thermal conductivity, but other explanations are possible. Within the central region of structural changes (Fig. 3–24), the PuO_2 particles could no longer be distinguished metallographically, but autoradiography revealed that the segregation had persisted almost unchanged. This principle could well be applied in other systems.

In general, the behaviour of $(U, Pu)O_2$ is similar to that of pure UO_2, so that more accurate measurements, or more closely controlled comparisons, will be needed to detect any differences.

ThO_2–PuO_2

Specimens of ThO_2 containing 2 to 19 wt% PuO_2 in solid solution were prepared by sintering (Freshley and Mattys, 1964). After exposures up to 3×10^{20} fissions/cm³ at power outputs as high as 1000 W/cm of fuel, the structures observed metallographically were similar to those seen in UO_2 or ThO_2–UO_2. Autoradiography showed that fission-product segregation occurred in these fuels too.

Al_2O_3–UO_2

Al_2O_3 and UO_2 form a pseudo-binary system with a simple eutectic at approximately 1900°C and 75 mol%

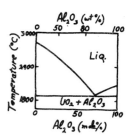

(a)

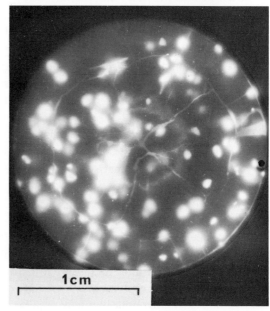

(b)

Fig. 3–24. Sintered UO$_2$–PuO$_2$ after irradiation: In fabrication the PuO$_2$ had been deliberately segregated.
(a) Micrograph.
(b) Autoradigraph — white indicates activity and, hence, plutonium concentrations.
(Courtesy A. S. Bain, R. D. MacDonald, and A. D. Murray, Chalk River Nuclear Laboratories.)

Al_2O_3. Specimens are normally prepared by compacting and sintering mixtures of the constituent powders. A negligible mutual solid solubility results in the Al_2O_3-rich compositions consisting of UO_2 particles dispersed in a matrix of Al_2O_3.

As part of their investigation of compartmented plate elements, Bleiberg et al. (1961) irradiated Al_2O_3–21wt%UO_2 (9 mol%UO_2) at fuel temperatures around 500°C and in cooling water pressurized to 150 atm. All three specimens increased in volume by about 20%, as did four other specimens of the same composition irradiated under quite similar conditions by Berman et al. (1960). The full increase had occurred by the lowest exposure tested, 9×10^{19} fissions/cm^3, and was unchanged up to 1.1×10^{21} fissions/cm^3. Metallography showed that most of the original porosity had disappeared as a result of the irradiation, so the true volume increase was probably nearer 30%.

In a specimen irradiated to 4×10^{20} fissions/cm^3, by which exposure about one fifth of the uranium atoms had fissioned, no grain boundaries could be observed in either the Al_2O_3 or the UO_2, nor could diffraction patterns for either phase be detected by x-ray studies. The larger UO_2 particles remained distinguishable, but they contained many pores of a few microns diameter. The same specimen had released about one percent of its fission-product krypton within the sheath during irradiation. Berman et al. had found, in specimens irradiated at temperatures under 75°C, that the crystalline structure of the Al_2O_3 had already been destroyed by an exposure of 10^{16} fissions/cm^3. The volume change on irradiation was therefore attributed to the Al_2O_3 being rendered amorphous. Williams (1963) reported that other specimens of Al_2O_3–UO_2 irradiated at similar low temperatures still showed diffraction lines for Al_2O_3 after 2×10^{17} fissions/cm^3: It is possible that in his specimens, which were not described in detail, the UO_2 was more coarsely dispersed, thereby damaging less of the matrix (Sec. 5.1).

From their results, Berman et al. calculated that a

volume of Al_2O_3 containing at least 10^6 atoms is affected
by each fission event, and they suspected that internal
stresses resulting from anisotropic thermal expansion
around the fission spike contribute to the lattice's instability. A similar phenomenon has been observed in U_3O_8
(Sec. 3.10), U_3Si (Sec. 4.7), and $ZrSiO_4$: Bombardment of
the Al_2O_3 and $ZrSiO_4$ by neutrons alone had no such
effect. All these structures that appear unstable are anisotropic, which supports the internal-stress hypothesis.

Although the large volume changes on irradiation can
obviously be a disadvantage to the use of Al_2O_3–UO_2 as
a reactor fuel, the results of Bleiberg et al. were, in one
regard, obtained under unfavourable conditions. Their
UO_2 particles averaged 5 µm in diameter so that virtually
all the Al_2O_3 matrix was exposed to damage by fission-
fragment recoils. The principle of increasing the size and
spacing of the fissionable particles, so that the majority
of the matrix remains undamaged, has been exploited in
other dispersion fuels (Sec. 5.1) and may well be applicable
here. Berman et al. observed that in a sample of irradiated
Al_2O_3–UO_2 the amorphous Al_2O_3 regained its crystalline
structure during an eight-hour anneal at 1000°C, while
Bleiberg et al. (1962) suggested that 800°C is sufficient to
cause redensification. The fuel would therefore probably
retain its crystalline structure if irradiated at a high enough
temperature. At lower temperatures the transformation
possibly depends on factors such as the rate of damage
and the external restraint.

BeO–UO₂ and BeO–(U,Th)O₂

The BeO–UO_2 system is similar to that of Al_2O_3–UO_2,
but the eutectic is at about 2100°C and 65 mol% BeO.
Bleiberg et al. (1961) included four elements containing
BeO–UO_2 with the other compartmented plates they
irradiated. Since all four were irradiated to approximately the same exposure, 1.1×10^{21} fissions/cm³, and
at the same temperature, around 500°C, the differences
in behaviour were attributable to differences in the original

structures. In one, containing 25.3 wt% UO_2 (9 vol%) as a fine dispersion of particles about 10-μm diam, the increase in volume was 27%, and 20% of the fission-product krypton had been released from the fuel. The UO_2 particles, which had undergone 54% fission of their uranium atoms, were full of pores a few microns in diameter, and were surrounded by much larger pores in a heavily damaged shell of BeO approximately 10-μm thick. Although not all the matrix was affected by these regions of heavy damage, presumably due to fission-fragment recoils, x-ray studies provided no evidence of BeO diffraction patterns.

At the other extreme, one specimen increased only 4% in volume and released 0.1% krypton. It, too, contained a fine dispersion of UO_2, but since the content was 65 wt% (33.5 vol%), the burnup of individual particles was down by a factor of nearly 4. Thus, the UO_2 particles still exhibited grain boundaries and were relatively free of porosity. There were no obvious signs of damage in the BeO matrix while diffraction patterns were obtained from both it and the UO_2. Intermediate in swelling and gas release were two specimens containing 28.7 and 30.1 wt% UO_2 (10 and 11 vol%) as coarse dispersions of particles about 150-μm diam. Although the burnup of individual particles was high, the greater size and separation of particles in these specimens resulted in less of the matrix being damaged, as judged by the remaining presence of BeO diffraction lines (Sec. 5.1).

Two further specimens, containing 51 and 65 wt% UO_2 as fine particles, were irradiated with estimated centre temperatures up to 1000°C. The increases in volume, about 26%, were larger than for the previous specimen of similar composition: The gas releases were 2.6 and 5%. At first the differences were attributed to a lower external pressure, since the last two had been irradiated in NaK-filled capsules (Bleiberg et al., 1961). Later, however, they were found to have achieved a much higher burnup (about 3.3×10^{21} fissions/cm³) than the previous one (Bleiberg et al., 1962). Thus, the effects of burnup, fuel temperature, and external pressure have not been properly separated.

Post-irradiation determinations of physical properties have shown that the BeO can be seriously affected by fission-fragment damage. In particular, the thermal conductivity is reduced, so that estimated fuel temperatures must be regarded as only nominal values. Gilbreath and Simpson (1958) measured the damage produced on irradiating specimens of BeO–2wt%UO$_2$ (0.55 vol%) and BeO–10wt%UO$_2$ (3 vol%) with surface temperatures estimated to have been about 250 and 600°C, respectively. After a burnup of only 5 × 10^{-6} of all atoms (7 × 10^{17} fissions/cm^3), linear dimensions had increased by nearly one percent, the thermal conductivity measured at temperatures below 100°C had decreased by a factor of 4 to 6, and both the compressive strength and elastic modulus had decreased by about 25%. Further exposure, by a factor of nearly 20, had little further effect on these properties. The damage began to be annealed at temperatures over 600°C. As in most materials, the structure-sensitive properties of BeO appear susceptible to considerable damage if the temperature is sufficiently reduced.

Other specimens, summarized by Williams (1963), irradiated at higher temperatures but to lower exposures than those of Bleiberg et al., suffered little obvious damage. For fuel in the range 800 to 1200°C, the volume increase probably rises gradually with exposure to a value about 4% at 5 × 10^{20} fissions/cm^3. At 2.4 × 20^{20} fissions/cm^3, there was no metallographic evidence of recoil damage in the BeO immediately surrounding the UO$_2$ particles. Pashos et al. (1964) reported less than 1% fission-product gas release from BeO containing 60wt% UO$_2$ (29vol%) that had been irradiated to 6 × 10^{20} fissions/cm^3 at an estimated centre temperature of 1180°C. In yet other specimens, containing 61wt% UO$_2$ (30 vol%) and irradiated to 3.7 × 10^{20} fissions/cm^3 at estimated temperatures up to 1310°C, the effect of UO$_2$ particle size was striking (Simnad, 1963). With coarse particles, about 150-μm diam, the pellets were intact, the volume had increased by 3% or less, but up to 25% of the fission-product gases had been released: With fine partices, about 20-μm diam, the pellets were cracked but had released less than 0.1 % of the fission-product gases. Presumably most of the fission fragments were retained by the UO_2 particles of the former, causing them to swell greatly and release gas, while in the latter the fragments largely escaped into the BeO matrix, embrittling it.

Hicks and Armstrong (1964) studied the release of fission products from BeO–10wt% UO_2 (3 vol%) by lightly irradiating specimens, then annealing them in the range 1225 to 1500°C. Appreciable amounts of molybdenum, tellurium, iodine, xenon, and cesium were released; also small amounts of yttrium, zirconium, barium, cerium, and neodymium. Elements in the second group could be just as mobile as those in the first, but just not sufficiently volatile to be released when they reached a free surface.

Since (U, Th)O_2 closely resembles UO_2, dispersions of (U, Th)O_2 in BeO would be expected to behave similarly to BeO–UO_2. Hanna et al. (1963) irradiated specimens containing up to 56 wt% (U, Th)O_2 (26 vol%) at measured centre temperatures from 600 to 850°C: The fissile phase ranged from uranium-rich in some to thorium-rich in others. Up to 1.4×10^{20} fissions/cm³, the increases in volume were 1% or less. There was a suggestion that fuel made from fissile particles of 100 to 180 μm was more stable than that using particles finer than 50 μm: X-ray studies showed that line broadening of the BeO reflections was more severe for the fine dispersions. Fission-gas release was under 1% for all specimens. Metallography revealed a tendency to weakening of the grain boundaries in the BeO, possibly due to internal stresses, but no other signs of damage.

Similar specimens were later irradiated to 2×10^{20} fissions/cm³ (Hanna and Hilditch, 1964). Again, dimensional changes were small with density decreases up to 1.4%. The fission-gas release from the fuel with coarse particles varied erratically from 0.1 to 5%, but that from the only two specimens with fine particles (under 20-μm diam) was less than 0.1%, just as found by Simnad for $BeO-UO_2$. Fissile particles that had suffered 5.5% burnup of the heavy atoms contained pores over 10-μm diam and were surrounded by a diffuse interface. Post-irradiation annealing at 1250 and 1500°C developed further pores in the particles and cracks in the matrix: The interface became sharp, but fine pores appeared in a layer of the matrix 15- to 20-μm thick around the particles.

ZrO_2-UO_2

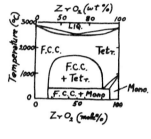

The crystal structure of the high-temperature modification of ZrO_2, although tetragonal, is similar to that of UO_2. Thus, solid solution of each in the other occurs, with extensive two-phase regions of solid solutions around the composition $ZrO_2-50wt\%UO_2$. The sketch provides a probable version of the equilibrium diagram, which is still the subject of controversy (Wright et al., 1964). The tetragonal/monoclinic transformation in ZrO_2 is suppressed by the addition of UO_2. Compositions around 50wt% UO_2, at temperatures below 1700°C, have a metallurgical structure consisting of some UO_2-rich grains and some ZrO_2-rich grains. However, the latter still contain approximately 30 wt% UO_2 so that their crystal lattice would be subject to recoil damage, even if the fissile-rich phase were segregated. In practice, the sintering process normally used in fabrication results in a grain size around 10 μm so that recoil damage is fairly uniformly distributed.

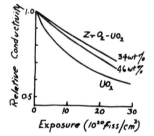

To determine the thermal conductivity, R. C. Daniel and Cohen (1964) irradiated sintered specimens of $ZrO_2-34wt\%UO_2$ (19 mol%) and $ZrO_2-46wt\%UO_2$ (28 mol%) containing thermocouples. For temperatures around 500°C and for exposures of the order of 10^{20} fissions/cm^3, they measured conductivities of 0.0165 and 0.021 W/(cm degC), respectively. The values were insensitive to temperature in the range 100 to 700°C, but decreased with exposure to 75 and 65% of the early conductivities by 2.5×10^{21} fissions/cm^3. Understandably, the ZrO_2 had lowered the conductivity of UO_2 markedly at low exposures, but by the time appreciable concentrations of fission products had been built up in both materials there was less difference in their conductivities. The conclusion regarding relative conductivities at negligible exposures was qualitatively confirmed by irradiations of very short duration (Bain, 1964). The mean thermal conductivity from 300°C to the melting point was deduced from the observed extent of central melting in sinters containing 60, 75, and 90 mol% UO_2 (76.7, 86.9, and 95.2wt%) as well as in pure UO_2. The addition of 40 mol% ZrO_2 had reduced the conductivity of the UO_2 by about 25%, and most of the decrease had already occurred by 10 mol% ZrO_2.

At high exposures, ZrO_2–UO_2 is susceptible to the same sort of volume increases already seen in UO_2 (Sec. 3.7), but there are some significant differences between the materials in their behaviour. Bleiberg et al. (1962) irradiated various compositions of ZrO_2–UO_2 containing from 18 to 94wt%UO_2 (10 to 87 mol%) as the fuel in compartmented plates. The fuel temperatures were estimated to range up to 1300°C, with the pressurized water coolant around 260°C. Unfortunately, the fuel/sheath temperature drop could not be calculated accurately since the fuel compartments had been evacuated on assembly. Thus, the heat-transfer coefficient would have been low initially, but would have increased markedly on the release of fission-product gases. The general absence of grain growth in the irradiated fuel indicates that the estimated temperatures had probably not been greatly exceeded.

The fuel behaviour with increasing exposure is illustrated in Fig. 3–25. Up to at least 3×10^{21} fissions/cm³ the fuel volume increases approximately linearly, reaching a value about 15% greater than the original volume. Initially, however, the fuel can decrease in volume: A decrease of as much as 7% was observed at an exposure under 6×10^{20} fissions/cm³. The macro- and micro-graphs of Fig. 3–25 show that the initial decrease results from collapse of the porosity originally present in the sinters. Consequently, the decrease in plate thickness is greatest in those elements originally containing fuel of low density. This initial densification has not been observed in UO_2.

The micrographs also show that fine spherical porosity develops in the ZrO_2–UO_2 at a much lower burnup than in UO_2, and that the progressive swelling is associated with an increase in the number and size of these pores.

There has been no observation of general breakaway in the swelling rates, such as occurred with some batches of UO_2, but some individual compartments were peculiar in exhibiting much greater swelling than the others in their plates. The lack of breakaway may be partly attributed to the fact that none of the ZrO_2–UO_2 specimens had compartment widths greater than 1/4 in., so that the fuel was moderately well restrained. Probably of greater significance was the fact that the fuel thickness was always under 0.05 in. for the ZrO_2–UO_2 plates. The fuel temperatures were thus maintained relatively low, even at high values of the surface heat flux and with the lower thermal conductivity of ZrO_2–UO_2.

By comparing the results from all their ZrO_2–UO_2 specimens, Bleiberg et al. concluded that swelling was slightly greater for those elements with estimated centre temperatures above 1000°C than for those at lower temperatures. The qualitative effect of temperature was clearly demonstrated in some of the compartments that bulged appreciably. On one side, the fuel pellet had become insulated from the sheath by a gas gap, with the result that the pores had enlarged to as much as 75-μm diam on that side: The restraint should have been closely similar on both sides. The composition of the ZrO_2–UO_2 specimens had no apparent effect on the swelling behaviour.

In the irradiated ZrO_2–UO_2, metallography failed to reveal any grain boundaries (Fig. 3–25), but x-ray diffraction proved that the material was still crystalline. With ZrO_2-rich compositions, Adam and Cox (1959) and Wittells and Sherrill (1959) found that the monoclinic phase was transformed to tetragonal by an exposure of 10^{16} to 10^{17} fissions/cm^3. Bleiberg et al. observed that all their specimens had become single phase by an exposure of 6×10^{20} fissions/cm^3 and supposed that fission spikes had homogenized the fuel in the same manner as has been proposed for uranium alloys (Sec. 2.6). The greater exposure required by the ceramic material is consistent with its original segregation having been on a coarser scale. By 10^{21} fissions/cm^3 the tetragonal phase had been transformed to cubic, and the latter remained up to 3.6×10^{21} fissions/cm^3. At these high exposures the concentration of fission products in the fuel may be contributing to the stability of the cubic phase. Where the fuel temperatures were high enough for pore growth to have become obvious a white precipitate similar to that attributed to segregated fission products in UO_2 was observed metallographically.

Bleiberg et al. measured the fission-product gas released from the fuel within compartments. The percentage release increased approximately linearly with exposure up to about 7% release at 3×10^{21} fissions/cm^3. Only those few compartments that exhibited unusual bulging had significantly higher releases, up to 67%. Unfortunately, there is no evidence to show whether the gas release caused the bulging, or *vice versa*: Whichever was first, the original cause of the instability is unknown. In post-irradiation annealing at 1000°C, Clayton (1962) found the gas release from high density ZrO_2–UO_2 was about 250 times that from comparable samples of UO_2. Thus, the earlier onset of observable porosity during swelling of the former may be due to more rapid migration of the fission-product gases, coupled with their having to migrate further to escape from the fuel when densification occurs. Bleiberg et al. suggested that the high degree of plasticity exhibited by the ZrO_2–UO_2 under irradiation may be associated with the anisotropic tetragonal phase, invoking the same mechanism proposed by Roberts and Cottrell to explain irradiation-induced creep of uranium (Sec. 2.4). In the hope of preventing the initial densification, Bleiberg and his colleagues are investigating the irradiation behaviour of ZrO_2–CaO–UO_2: Since CaO stabilizes the cubic phase in this system, anisotropy effects might be avoided.

MgO–PuO_2

Since MgO and PuO_2 are mutually immiscible in the solid state, a dispersion of PuO_2 particles in MgO can be formed.

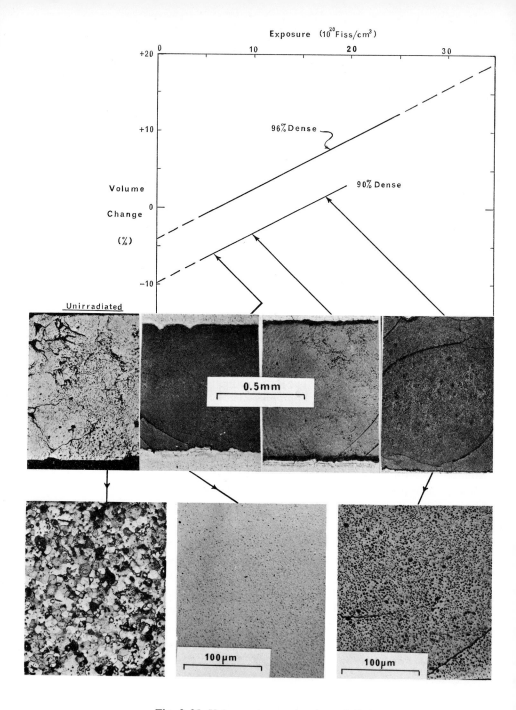

Fig. 3–25. Volume changes in sintered ZrO_2–UO_2 as a function of fuel exposure for two different initial densities, with micrographs showing the fuel structure at selected points. (After Bleiberg et al. (1962), courtesy International Atomic Energy Agency.)

Freshley and Carroll (1963) have conducted some exploratory irradiations of sintered specimens of MgO–2.71wt% PuO_2 and MgO–12.95wt%PuO_2 (0.9 and 4.7%vol%) whose densities ranged from 86 to 92% of that theoretically attainable. The fuel was heavily restrained by the sheath and was irradiated for exposures up to 10^{20} fissions/cm^3 (72% burnup of the plutonium) at estimated power outputs up to 1650 W/cm of fuel: From values for the thermal conductivity of unirradiated MgO, the centre temperatures were calculated to range from 700 to 2450°C. The cross sections resembled, in some aspects, those of irradiated UO_2 with the hottest specimens exhibiting central voids surrounded by broad and narrow columnar grains, then enlarged equiaxed grains, and as-sintered material. However, any given effect occurred at a higher power output in the MgO–PuO_2, and cracking was generally less prevalent, presumably because of a higher thermal conductivity than in UO_2.

Most striking was the observation that PuO_2 particles had largely disappeared from the translucent broad columnar grains, and, together with fission products, had concentrated in a band further out. In one specimen the segregation occurred between the two types of columnar grains, while in another it lay within the narrow columnar grains. Analyses for plutonium showed a tenfold enrichment over the original concentration in the segregated band, and a tenfold depletion in the broad columnar grains. These observations support the idea that the broad columnar grains can form by their grain boundaries moving outwards (Sec. 3.2), but details of the process remain obscure.

Although the PuO_2 particles were fine (probably less than 10 µm) the MgO crystal structure was undamaged by the exposure. The only change observed was in the band enriched in PuO_2 and fission products, where the intensity of the x-ray reflections was reduced by a factor of 3. The fission-product gas release from these porous sinters varied from 5 to 38%, with the greater releases apparently occurring from the hotter specimens.

Carbide Fuels and Other Fissionable Compounds

4.1 Uranium-Carbon System and Relevant Properties

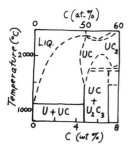

In the uranium-carbon system there are three compounds, UC, U_2C_3, and UC_2, but the wish for a high uranium density has led to the first having been studied most thoroughly. The preparation and properties of all three have been reviewed by Frost (1963). Stoichiometric uranium monocarbide occurs at 4.8 wt% carbon. At lower carbon content, say 4.4 to 4.8 wt%, uranium metal is present, usually at the grain boundaries. Although not obvious from the diagram, UC has a finite solubility for uranium; a maximum of 1.5 at.% (0.25 wt%) at about 2000°C, diminishing to zero at both the melting point (about 2400°C) and the eutectic temperature (1117°C). Specimens in this composition range can retain the excess uranium in solid solution when cooled rapidly, but, when they are cooled more slowly, the uranium precipitates within the grains as sub-micron-sized cubic particles. Hyperstoichiometric UC, say 4.8 to 5.2 wt% carbon, often exhibits a Widmanstätten structures of UC_2 platelets in the UC grains: At temperatures below 1800°C, the UC_2 is metastable, but conversion to U_2C_3 does not occur easily below 1000°C.

The monocarbide has a face-centred cubic structure of the NaCl type. Excess uranium in solid solution decreases the lattice parameter, probably by the formation of carbon vacancies. In contrast to UO_2, in which ionic bonding probably predominates, UC contains covalent and metallic

○=C ●=U

bonds (Frost, 1963). In U_2C_3 and UC_2, the U-C bonds are weaker, and adjacent carbon atoms are covalently bonded. U_2C_3 has a body-centred cubic structure with eight formula units per unit cell: UC_2 can be either body-centred tetragonal or face-centred cubic (CaF_2-type), with the former being the equilibrium structure above 1800°C. The electrical and thermal conductivities of UC are lower than those of most metals. However, the ratio of the thermal to the electrical conductivity is proportional to the absolute temperature (the Wiedemann-Franz Law for metals), which indicates that conduction by electrons predominates. The other two carbides have lower conductivities but are still in the range of semi-metals.

As in the case of UO_2, uranium carbides can be fabricated by compacting and sintering powders, with varying amounts of porosity remaining in the product. Reaction-sintering the component elements is also employed, but this can result in nonequilibrium structures with uranium, UC, and UC_2 coexisting. The most widely used method of preparation is arc-melting and casting. Its product is susceptible to porosity, impurity content (mainly oxygen and nitrogen), microcracking, and inhomogeneity of composition, but these defects can be avoided in a well-controlled process.

4.2 Structural Changes

Irradiated specimens of uranium carbide often exhibit significantly different structures from those of the unirradiated material, but it is not always clear how much of the change is truly due to the irradiation. To aid heat transfer to the sheath, the carbide is commonly irradiated while immersed in liquid sodium or NaK. For examination, the liquid metal is dissolved out or evaporated as completely as possible. However, any traces left within the body corrode on exposure to the atmosphere, causing cracks to propagate. Furthermore, even clean uranium carbide is susceptible to atmospheric attack on standing at room temperature. For both reasons, specimens that appeared relatively intact on first post-irradiation inspection have been extensively cracked a few days later. The appearance of microcracks can also result from pull-out during polishing for metallographic examination. Any crack observed during examination must therefore be treated with suspicion, but some gross cracking of the carbide probably does occur during irradiation, especially where large thermal differences exist in the fuel.

The self-diffusion of carbon in UC is relatively rapid, but thermal gradients alone do not cause segregation of UC_2 precipitates. In a laboratory test, a specimen of $(UC + UC_2)$ was held in a thermal gradient of 500 deg C/cm at temperatures up to 1900°C for 93 h without any appearance of UC_2 migration in either direction (Osthagen and Bauer, 1964). However, the presence of a liquid metal can again complicate matters since the solubility of carbon in sodium, although low, is sufficient to permit mass transfer. If the sheath is of a material that can provide a sink for the carbon, e.g., steel, UC_2 in the specimen is converted to UC by loss of carbon (Hahn, 1963). After irradiation under such conditions, specimens that were originally hyperstoichiometric uranium monocarbide may no longer contain precipitated dicarbide particles. Sometimes the precipitate left is confined to an outer cool rim of the fuel, suggesting that the carbon first diffuses out through hot surfaces such as cracks, pellet end faces, or thermocouple holes, where these are present (Sinizer et al., 1962). Conversion of a UC_2 platelet to UC may also result in formation of a microcrack, thereby helping the liquid metal reach more UC_2. Some other sheathing materials, e.g., niobium, form thin protective films of refractory carbides, which can prevent significant leaching of the carbon from the fuel.

In hypostoichiometric specimens that have operated with temperatures above 1000°C, the free uranium, originally dispersed uniformly, has been observed to have migrated outwards from the hot centre (Crane et al., 1964). Since the central region is in compression during operation and the uranium is present either as a liquid or the soft gamma phase, the free metal may be simply squeezed along the grain boundaries. The thermal expansion of the uranium, its volume change on melting, and irradiation-induced swelling of the uranium would all help to expel the metal if it formed a continuous phase at the grain boundaries.

Isothermal grain growth occurs in unirradiated UC at temperatures around 1500°C, just as it does in UO_2. However, few carbide specimens have been irradiated at temperatures over 1300°C, so there is little useful experience of grain growth in UC during irradiation. In some specimens whose sheaths ruptured during irradiation the UC may have melted, but means of identifying UC that solidified during irradiation have not been established.

Using electron microscopy, Frost et al. (1962a) showed a striking difference in the appearance of the fractured surface of UC irradiated at 60°C when the exposure* exceeded 10^{16} fissions/cm³. The transgranular fracture of unirradiated material became intergranular and holes or cracks were apparent at grain boundaries. Similar observations have been made on the fractures of uranium (Sec. 2.9) and UO_2 (Sec. 3.2). However, the carbide was a reaction-sintered product in which free uranium may have existed. Further experiments on a well-characterized sample of uranium carbide are desirable.

For some time the only observations of structural changes really due to irradiation were associated with uranium or graphite present in the carbide. Bradbury et al. (1963) observed very fine pores in the uranium phase in grain boundaries of irradiated hypostoichiometric mono-carbide (Fig. 4–1). In this respect the uranium was behaving in the same manner as it does in bulk form. When these pores grow larger, they occupy the full width of the uranium layer between adjacent UC grains. Bradbury et al. also noted very fine pores and precipitates, probably UC_2, in a thin layer of UC immediately adjacent to a graphite

---

* In UC, 1 % uranium burnup $\equiv 3.3 \times 10^{20}$ fissions/cm³.

inclusion. Freas et al. (1961) found that the outline of UC_2 platelets became blurred during irradiation, and they showed electron micrographs of tiny nodules in the UC that had been depleted of UC_2 by the liquid metal. From the etching behaviour, they tentatively identified the precipitate as graphite, but in their specimens with around 1% burnup fission-product carbides might have formed. More recently, Crane et al. (1964) have shown irradiation-induced porosity in the UC phase of cast specimens exposed to about 5% burnup of the uranium at temperatures over 1000°C. Their observation that the pores occurred preferentially at the hotter side of each grain was of great interest: This suggests pore migration up a thermal gradient as seen in unirradiated UO_2 (Sec. 3.6).

4.3 Conductivities and Density

In contrast to the minor effects of irradiation on the appearance of uranium-carbide fuel, very large changes have been observed in the electrical resistivity and crystallographic cell size of some specimens. These experiments have been intended to study irradiation damage: Thus, the specimens were deliberately irradiated at low temperatures (generally below 100°C) to retain damage, and the properties selected were those most likely to reveal lattice damage. Accurate density measurements have also shown changes due to irradiation. Several parallels can be seen to the effects of irradiation in UO_2, and in neither material is the nature of the damage understood in detail.

Childs and his colleagues (1962, 1963a, 1963b) measured the electrical resistivity at 20°C of cast uranium carbide with a large grain size (a few hundred microns). They found (Fig 4–2a) a large increase of about 150% in resistivity, which saturated at exposures around 6×10^{16} fissions/cm^3. A few measurements made at temperatures down to -196°C demonstrated that the increment in resistivity due to irradiation was independent of the measuring temperature, which suggests that the effect resulted from a high concentration of point defects. The inability to fit the experimental results to a simple exponential dependence of damage on exposure indicated that more than one type of defect was involved. The values obtained by Freas et al. (1961) for cast uranium carbide at much higher exposure (Fig. 4–2a) are consistent with the others: Although their resistivity increases were appreciably larger, the inevitable difference in specimen fabrication could well be responsible.

The resistivity increment at saturation was shown by Childs et al. to depend on the stoichiometry and impurity content of the specimens. Hypostoichiometric samples exhibited lower increments, down to 90% of the unirradiated value, but the correlation with the analysed composition was poor. Presumably that part of the excess uranium that was precipitated as free metal was not contributing. The correlation was very much better between the saturation increment and the cell size of the unirradiated material, which is a measure of the concentration of carbon vacancies in the UC structure (Sec. 4.1). Annealing hypostoichiometric specimens prior to irradiation, to precipitate the excess uranium, increased the saturation increment as expected. However, the increment was then greater than that of stoichiometric specimens: This suggests that certain defects present in the unirradiated material could annihilate defects caused by irradiation. The presence of tungsten impurities in the UC also reduced the saturation increment.

Increases in electrical resistivity were observed by Griffiths (1961) in uranium carbide prepared by reaction-sintering of the component elements, but the saturation increment was less than 3% of the unirradiated resistivity. The difference could have been partly the result of a greater concentration of impurity atoms in the sintered material (oxygen and nitrogen contents would be expected to be higher), but Griffiths (1963) has demonstrated that variations in grain size were primarily responsible. As-sintered, the uranium carbide was composed of grains 10- to 20-μm diam: Pre-annealed specimens of 40- to 45-μm

206

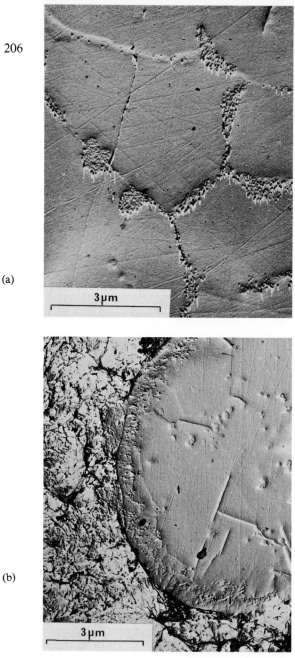

(a)

3 µm

(b)

3 µm

Fig. 4–1. Microstructures in irradiated uranium carbide.
(a) Electron micrograph showing porosity in the uranium phase
at grain boundaries in hypostoichiometric UC.
(b) Electron micrograph showing porosity and precipitates in UC
phase near graphite inclusion. (a & b after Bradbury et al. (1963),
courtesy Macmillan and Co. Ltd.)

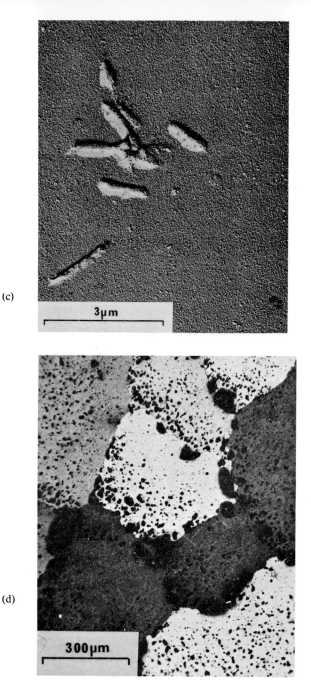

(c) Electron micrograph showing nodules that appeared in hyper-stoichiometric UC. (After Freas et al. (1961), courtesy ASTM.)

(d) Micrograph showing extensive porosity in near-stoichiometric UC: Note tendency for pores to be concentrated at the same side of each grain. (After Crane et al. (1964), courtesy AIME.)

grain size developed saturation increments of nearly 60%. The explanation could be that grain boundaries act as sinks for point defects, that the number of point defects produced by a fission fragment is greater in a perfect lattice, or that a point defect at a boundary has less effect on the resistivity.

The irradiation damage to the electrical resistivity of uranium carbide can be removed by thermal annealing (Fig. 4–3a). Initially, Childs et al. found that about 70% of the damage in stoichiometric material recovered about 150°C and the remainder about 500°C. Increasing exposure and impurities tended to smear the recovery stages, which still occurred at roughly the same temperatures. Later work revealed a further 1.3% recovery at 1200°C, with some residual damage remaining even after five hours at 1500°C. A hypostoichiometric specimen behaved similarly up to 500°C, but the resistivity then decreased below the unirradiated value and passed through a minimum about 700°C: The reproducibility of this behaviour has not yet been demonstrated. Damage in the fine-grained sintered material also showed partial recovery near 150°C. However, the only other stage observed was about 700°C where approximately 60% of the recovery occurred. The comparison probably reflects a variation in the proportion of different types of defect in the two materials.

Increases in the lattice parameter, shown in Fig. 4–2b, also exhibited saturation with increasing exposure (Childs et al.). Once more, the results of Freas et al. at much higher exposures were in agreement. Excess uranium and tungsten impurities again reduced the magnitude of the saturation damage, presumably by introducing carbon vacancies, with the amount of reduction depending on the composition. Prior annealing of hypostoichiometric specimens caused an increase in the saturation increment in lattice parameter, but the resulting value was lower than that for stoichiometric specimens (cf, the effect on resistivity). This behaviour could be the consequence of a higher ratio of vacancies-to-interstitials being produced in the annealed lattice: Both defects would increase the resistivity but only interstitials would decrease the lattice parameter. Density changes plotted in Fig. 4–2c, although very small, confirmed the increases in lattice parameter, which indicated that any cracks or porosity formed in these specimens occupy a negligible volume (Childs and Buxton, 1965). In hypostoichiometric specimens the lattice parameter resumed its increase at exposures over 10^{18} fissions/cm^3 and an increase in density was observed, but in only one specimen. These observations suggested that precipitation of free uranium was starting at high exposures, although the temperature was very much lower than that at which uranium would precipitate in unirradiated specimens.

Results on annealing an irradiated stoichiometric specimen were not sufficiently extensive to locate precisely the recovery stages, but one was between 0 and 400°C, while another between 400 and 600°C fully restored the lattice parameter (Fig. 4–3b). Recovery of the density decrease was generally similar until over 1200°C, when the density increased beyond its pre-irradiation value (Fig. 4–3c). Thus, this behaviour is consistent with the annealing of resistivity damage.

The annealing behaviour of hypostoichiometric specimens was complex. In the one illustrated in Fig. 4–3b, the lattice parameter nearly recovered its original value at 300°C, then rose again to a maximum about 500°C, before finally recovering by 700°C. The maximum approximately coincided with a minimum in the breadth of the x-ray diffraction line. In another specimen, irradiated to a lower exposure, the minimum line breadth occurred at about the same temperature, but the maximum in the lattice parameter was at approximately 300°C. The density decreases of two specimens recovered partially by 400°C, then their densities increased rapidly beyond the unirradiated values, passed through a maximum about 800°C, and by 1500°C had dropped below the as-irradiated values. In another specimen, the density had regained its pre-irradiation value by 300°C, then passed through a maximum

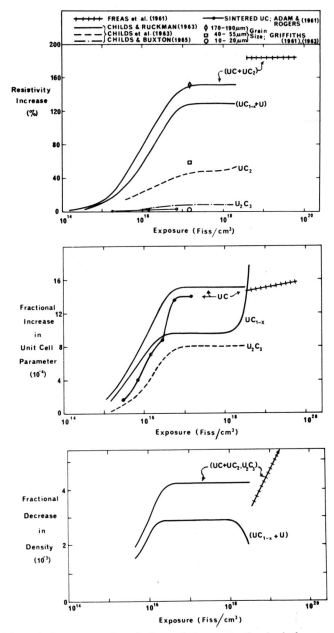

Fig. 4–2. Some examples of observed damage to the physical properties of uranium carbides as a function of the irradiation exposure. (a) Electrical resistivity. (b) Unit cell parameter. (c) Density.

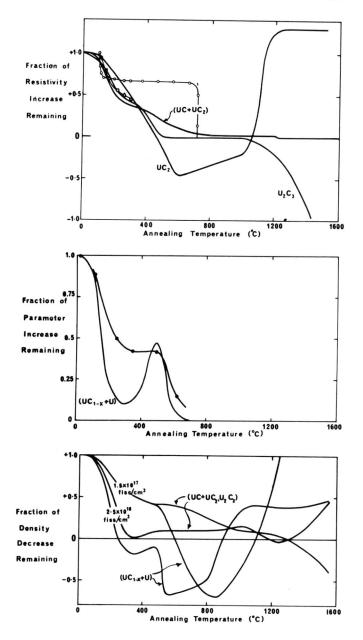

Fig. 4–3. Some examples of observed annealing of irradiation damage in uranium carbides. (Key to symbols as in Fig. 4–2.)

(a) Electrical resistivity.   (b) Unit cell parameter.   (c) Density.

between 500 and 800°C before again decreasing up to 1500°C.

Childs and Buxton (1965) suggested that the annealing of density decreases depended more on exposure than on composition. In the first stage of recovery, the density change was much greater for specimens irradiated to 2×10^{18} fissions/cm^3 than for those irradiated to only 10^{17} fissions/cm^3. They speculated that this stage represented agglomeration of uranium interstitials onto platelets, in a manner analogous to that for lithium in irradiated lithium fluoride, which has the same crystal structure as uranium monocarbide. The densification on annealing hypostoichiometric specimens at higher temperatures was attributed to precipitation of free uranium, but, without any metallographic confirmation. The final decrease in density may have been due to cracking induced by oxidation of the uranium, since there was no corresponding change in the lattice parameter, and unirradiated specimens exhibited similar decreases in density at these temperatures.

The relatively sharp recovery stages in the annealing of resistivity damage suggest that specific point defects are becoming mobile at these low temperatures, possibly to form defect clusters. Such clusters would be expected to have less effect than the point defects on the lattice parameter or electrical resistivity, so their further development might not have been observed in the present experiments. Complementary studies by electron microscopy are obviously desirable. The annealing of the change in lattice parameter in hypostoichiometric material reveals a more complex situation, confused by the possibilities of uranium precipitation or dissolution and dimensional changes of the free uranium phase. Thus, there is no consistent explanation of the damage behaviour. The pre-annealed specimens indicated that at least two types of defect were being produced in varying ratio, and Childs et al. argued that these were more likely to be uranium vacancies and interstitials than carbon ones. Frost et al. (1962a) showed by quenching experiments in the laboratory that vacancies (either uranium or carbon) become mobile about 700°C:

Therefore, they attributed the recovery stages at lower temperatures to interstitial movement.

Contrary to the situation with electrical resistivity, the increase in lattice parameter with irradiation was approximately the same in reaction-sintered material as in cast uranium carbide. Adam and Rogers (1963) concluded that, in the sintered product, saturation damage in the resistivity at about 10^{16} fissions/cm^3 coincided with a first stage of damage in the lattice parameter. Saturation did not occur in the cast uranium carbide until an exposure approximately four times higher. They further identified the fraction of damage in each stage with the fraction recovered in each of the two annealing stages, which occurred at the same temperatures as for stoichiometric cast UC. However, their suggestion of two stages in the damage is based on slender evidence and must be regarded as an intriguing possibility until the reproducibility of the effect is established. Figure 4–2b suggests another possibility; that stoichiometric and hypostoichiometric UC coexisted in the sintered material.

Irradiation caused an increase in the resistivity of UC$_2$ by about 50% of the unirradiated value at 1.5×10^{18} fissions/cm^3, but the damage may not have saturated by this exposure (Childs et al., 1963b). Annealing restored the resistivity to its initial value by 500°C without any discrete recovery stage. The resistivity then passed through a minimum and rose to a high value above 1000°C. Removal of pre-existing imperfections and partial transformation to the stable (U$_2$C$_3$ + C) probably caused the high temperature effects.

Irradiation damage to the resistivity of U$_2$C$_3$ was very similar to that of UC, except that saturation occurred at an increase of only 5%. The annealing behaviours were also similar. However, the U$_2$C$_3$ studied by Childs et al. was of small grain size, and, by analogy to UC, this fact may have reduced the damage. Freas et al. (1961) observed a 250% increase in the resistivity of U$_2$C$_3$ of unspecified grain size, although, admittedly at a higher exposure. Childs et al. found an expansion in the lattice, while Freas et al. found a contraction. Thus, in this material also, the ratio of different types of defect formed may depend on the degree of perfection of the original lattice.

Very little is known concerning the effect of irradiation on the thermal conductivity of uranium carbides. Rough and Chubb (1960), from irradiation tests with thermocouples in uranium carbide at temperatures from 300 to 800°C, deduced in-reactor conductivities marginally lower than that for unirradiated material. However, in view of the experimental errors, they considered that any apparent difference was of doubtful significance. Their observations extended to a maximum exposure of 5.5×10^{20} fissions/cm^3.

Since unirradiated uranium carbide obeys the Wiedemann-Franz Law, the thermal conductivity presumably suffers damage similar to that in the electrical conductivity. The thermal measurements were made on specimens intended more to simulate fuel elements for power reactors than to retain damage for study. The much smaller damage experienced by the thermal conductivity than by the electrical conductivity may therefore be explained simply by the higher fuel temperatures for the former. Annealing of the damage to the electrical conductivity of uranium carbide (Fig. 4–3a) indicates that irradiation temperatures above and below 150°C should give widely differing results: A similar effect might also be expected around 500°C. Further experimental studies on the effect of irradiation temperature on both conductivities are required, as it has been shown that serious damage can occur at low temperatures.

4.4 Dimensional Stability — Cracking and Swelling

The dimensional changes produced on irradiating uranium carbide are affected by several factors and are due to a variety of mechanisms. They have been studied by measuring changes in the fuel's diameter and in its displacement density and by metallography. Sinizer et al. (1962) analyzed their experimental results by plotting the percentage increase in diameter per 1% burnup of the uranium as a function of the fuel's centre temperature: Figure 4–4 is a similar display prepared from more recent results. There is evidence that the diameter increase rises at higher temperatures.

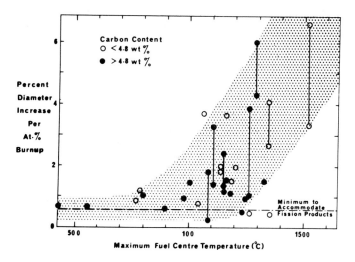

Fig. 4–4. The increase in diameter of UC per unit burnup as a function of the irradiation temperature. Derived from the results of Hahn (1964) and Crane et al. (1964).

A series of experiments, due to Crane et al. (1964), has demonstrated very clearly the importance of fuel composition in the temperature range where expansions are appreciable (Fig. 4–5a). The estimated centre temperatures of these specimens were normally in the range 1000 to 1200°C, but at times some reached as high as 1350°C. Thus, parts of almost all the specimens would at some time have exceeded the melting point of uranium (1130°C). Two of the specimens exposed to the highest burnup (about 5% of the uranium) showed relatively small density decreases *per 1% burnup*, even though their actual decreases were over 20%. The liquid used in the displacement method probably penetrated the larger pores, so that the

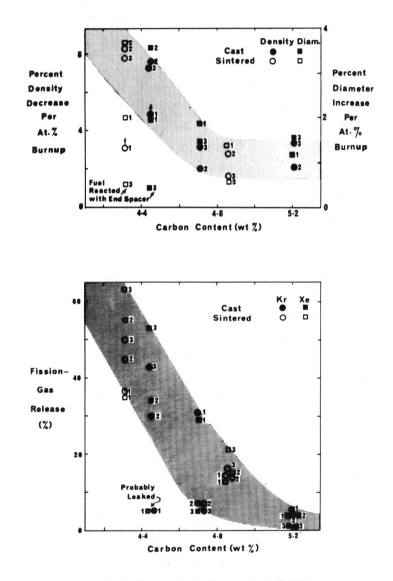

Fig. 4–5. Effect of the carbon content of UC on
(a) the density decrease per unit burnup,
(b) the fission-gas release.
Derived from the results of Crane et al. (1964); revised values for xenon obtained as a private communication from the authors.

density decreases measured are lower limits. Apart from this qualification, the results plotted in this manner do not exhibit any marked dependence on the burnup, which varied by a factor of 3, thereby providing some justification for normalizing dimensional changes to 1% burnup.

The same authors compared the behaviour of cast and sintered specimens but found that any differences in the density decreases were explained by compositional differences without any apparent effect of the fabrication route (Fig. 4–5a). They also compared helium-filled, low-clearance specimens with others in which the fuel was immersed in liquid sodium. Although there was a suspicion of greater cracking in those filled with sodium, there was no obvious effect on the density decrease.

Some of the observed density decreases are due simply to fission products in the lattice. Rough and Chubb (1960) estimated an upper value for the expansion of 1.6% density decrease per 1% burnup: This reduces to 1.0%, if fission-product atoms can reach vacant uranium sites. Swelling due to fine porosity, presumably filled with gaseous fission products, has been observed in UC (Sec. 4.2) but only in material that had been exposed to 5% burnup at central temperatures up to 1150°C. At lower temperatures, Fig. 4–4 suggests that some other mechanism is operative.

The increase in fuel diameter would equal approximately one third of the density decrease, if the expansion were uniform and isotropic, or one half, if there were no elongation. However, Hahn (1963) pointed out that the increase in diameter was often much larger than this and could vary irregularly along the length of a fuel element. In one particular element, the value of one third the density decrease lay between 0.4 and 1.3%, but the diameter increase ranged from 0.5 to 5.0%. He attributed the difference to cracking. Where a crack develops in the fuel but is arrested partway through so that fragmentation does not occur, the diameter normal to the crack is increased without affecting the displacement density. The low incidence of large cracks explains the variability in the diameter measurements.

By selecting empirically the factor 2.4 between diameter increase and density decrease, Crane et al. obtained good agreement between the two methods for all their specimens (Fig. 4–5a). This consistency and the fact that 2.4 lies comfortably between the values 2 and 3 that can be justified theoretically indicate that cracking contributes little to increases in diameter of their specimens. In the gas-filled specimens the tight-fitting sheath restrained the fuel: Even in the sodium-filled specimens, the fuel was held centrally in its sheath by a spiral of molybdenum wire. On examination, the sheaths were found to have expanded slightly, while the wires had become embedded in the fuel. Thus, in both these types of specimen, the fuel was subjected to some restraint. Many of the other uranium-carbide specimens have been irradiated while immersed in liquid metal but without spacer wires. Under such circumstances, cracks in the fuel would be free to open up without meeting external resistance. Even a weak sheath might be able to stop the fuel's diametral expansion due to cracking, but the volume decrease due to the generation of fission products and fine porosity would probably be much more difficult to restrain.

Figure 4–5a shows that the changes are much greater in hypostoichiometric material: The volume increase is associated with the uranium phase (Sec. 4.2). Fine pores nucleate in the uranium, then grow to occupy the full width of the grain-boundary layer. In addition, outwards migration of the free uranium may cause porosity in the central region. In hyperstoichiometric specimens immersed in liquid metal, leaching of the carbon from UC_2 platelets can, by the formation of microcracks (Sec. 4.2), contribute to the density decrease. However, initially, hyperstoichiometric material is sometimes rendered single phase without microcracks developing.

The existence of several mechanisms for expansion in uranium carbide makes it difficult and uncertain to predict the fuel's behaviour. For instance, expansion due to the solid fission products should increase with burnup but be independent of temperature; swelling due to gaseous fission products should increase with both burnup and temperature; and increase in diameter due to cracking should be independent of burnup and depend mainly on the temperature *difference* in the fuel, although it may decrease slightly with the mean temperature as the fuel becomes less brittle. Despite these complications, it has been established that a burnup of 5% of the uranium can be achieved at temperatures up to 1150°C with a density decrease of 15%. Under these conditions, porosity in the UC phase is apparent and swelling would be expected to pose a serious problem at higher temperatures or higher exposures.

4.5 Fission-Product Gas Release

Although observations of fission-product gas release from UC are less extensive than from UO_2 (Sec. 3.6), the two materials exhibit some similarity. Most measurements of the diffusion coefficient of xenon in UC have used the technique of a light irradiation at low temperatures followed by a diffusion anneal. In early determinations, Lindner and Matzke (1959) obtained values appreciably above those for UO_2, at temperatures around 1100°C: Auskern and Osawa (1962) obtained values lower than those for UO_2 in an overlapping temperature range. Other determinations, including one made during irradiation (Melehan and Gates, 1964), yielded values between these two extremes (Fig. 4–6).

Reconciliation of all the results is not yet in sight but some reasons for the variability are now apparent. Matzke (1962) reported that oxygen and nitrogen impurities in the UC decreased the activation energy for xenon diffusion, and later measurements of the diffusion coefficient by Matzke and Lindner (1964) show better agreement with the others. Another member of the same group showed that the apparent diffusion coefficient depended on the prior exposure of the specimen: Below 10^{15} fissions/cm^3, the

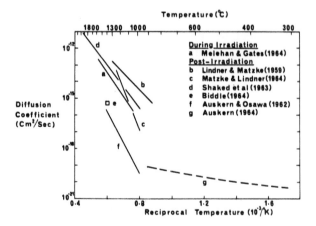

Fig. 4–6. Dependence on temperature of the diffusion coefficient for fission-product gases in UC.

results agreed well with those of Shaked et al. (1963) (Fig. 4–6): At 2×20^{17} fissions/cm³, they were close to those of Auskern and Osawa (Papathanassopoulos, 1963). The specimens of Auskern and Osawa were irradiated to a higher exposure than those of Shaked et al., but the difference in exposure (2.5×10^{15} fissions/cm³ for the former versus 8.5×10^{13} to 4×10^{14} fissions/cm³ for the latter) does not appear sufficient to account for the disagreement. By analogy to the work of MacEwan and Stevens with UO_2 (Sec. 3.6), the dependence on exposure of the xenon release from UC can be attributed to internal trapping of the gas. If so, migration of gas-filled pores may be an important mechanism for gas release.

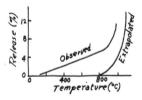

The xenon release from nearly stoichiometric UC at low temperatures has been measured by Auskern (1964). In the range 150 to 900°C, the gas collected in a given time increased approximately linearly with temperature: Although the amounts were small, they were considerably larger than would have been predicted by extrapolation to lower temperatures of the earlier results by Auskern and Osawa. If Auskern's low-temperature releases were analysed in terms of normal bulk diffusion, the interrupted line of Fig. 4–6 would be obtained. However, he argued that the results were due to release from a surface layer (roughly three or four unit cells deep) occurring over a range of activation energies lower than that for bulk diffusion. A possible interpretation is that for the first few jumps of a xenon atom in the region damaged by a fission spike vacancies do not have to be supplied, so the activation energy for diffusion is reduced. Alternatively, localized disorder of the lattice structure in the layer close to a free surface may reduce the activation energy.

Comparative measurements (Stevens, 1963) on specimens of four different carbon contents from 4.4 to 6.0 wt% suggested that increasing the carbon content in this range substantially reduced the xenon release at 1400°C. Although his specimens were all prepared by arc-melting, their specific surface areas were not measured, so the dependence of release on composition was not conclusively established.

Post-irradiation analyses of the gas released from uranium carbide and collected within the sheath are also informative. Many releases have been under one percent of that generated and have come from fuel operating at temperatures under 1000°C. The mechanism for these small releases is probably the easy diffusion from surface layers noted by Auskern in his post-irradiation annealing studies. However, recoil of fission fragments into a liquid metal and ejection from fission-fragment tracks that intersect the surface (Sec. 3.6) may also be contributing.

Releases as high as 50% have been found for some specimens with maximum fuel temperatures up to 1350°C. The experiments of Crane et al. (1964), already mentioned in connection with dimensional changes (Sec. 4.4), are most valuable in showing how strongly the release depends on the amount of free uranium present (Fig. 4–5b). Since the percentage gas release is much greater than the percentage free uranium in the specimen, the uranium phase, or porosity in it, appears to act as a network of easy-escape paths for gas from the UC phase as well. The similarity between the plots of density decrease and gas release (Figs. 4–5 a & b) suggests that it is swelling of the uranium phase that causes the gas release, and not retention of the gas in the UC phase that causes the majority of the dimensional changes. Generally, the *percentage* release did not change as the burnup increased by a factor of more than 2. For the specimens containing 4.7 wt% carbon, however, there was a greatly increased release at the highest burnup. The authors associated this observation with an increase in the amount of free uranium in the same specimen, seen during post-irradiation metallography. The possibility that fission products are removing carbon from the uranium is interesting and merits further study: However, in this experiment, the sheath or centering spring could be responsible for removing carbon through the liquid metal.

For compositions near stoichiometric, there was an indication that sinters, with their greater surface area, were releasing more gas than cast specimens, at least for exposures up to 3 at.% burnup. Although this series provides valuable internal comparisons, it is difficult to compare the absolute magnitudes of the release with those from other tests. This is because the specimens used were very small (3/8-in. diam by 3/4-in. long), so that the distribution of burnup and temperature is hard to estimate.

Figure 4–7 provides a general correlation between observed gas releases and fuel temperatures. The results indicate that the release can be appreciable at temperatures over 1000°C, and a rapid increase with increasing temperature in this range might be expected from analogy to UO_2 (Sec. 3.6). A direct comparison between results for the two materials suggests that UO_2 has the better gas retention, but such a conclusion could be misleading for several reasons. In Fig. 4–7 the specimens showing greatest release are hypostoichiometric: Also, the surface temperature and burnup of carbide specimens for which gas-release data are available have generally been higher than those of the corresponding oxide specimens. The very large scatter of results illustrated in Fig. 4–7 emphasizes the need for well-controlled experiments designed to determine the effect of temperature on the release of gas from carbide specimens of specific composition. The analogy to UO_2

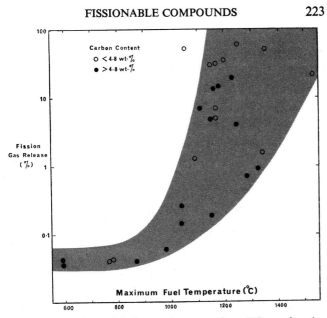

Fig. 4–7. The release of fission-product gases from UC as a function of the irradiation temperature. Derived from the results of Hahn (1964) and Crane et al. (1964).

could be inappropriate, if the migration of gas-filled pores is an important mechanism for gas release from UO_2: The higher thermal conductivity of UC results in lower thermal gradients and, hence, reduces the driving force for pore migration.

4.6 Other Carbides, Including Mixtures

Some qualitative observations on the irradiation behaviour of uranium dicarbide, mixed uranium and thorium dicarbide, and mixed uranium and zirconium monocarbide have resulted from the testing of coated particles (Sec. 5.5). Mixed uranium and zirconium monocarbide has also been successfully irradiated in thermionic diodes at temperatures approaching 2000°C but to exposures negligible compared with the requirements of reactor fuel. Otherwise, experience is confined to plutonium monocarbide and mixed uranium and plutonium monocarbide. Although the latter is probably the fuel with greatest potential for fast reactors, few specimens have been irradiated. There is mutual solubility between the two stoichiometric compounds UC and PuC: At 15% plutonium, a possible composition for fast-reactor fuel, any excess carbon associates with the plutonium to give (UC + Pu_2C_3), while a shortage of carbon results in a uranium-plutonium alloy being precipitated from the stoichiometric mixed (U, Pu)C.

Kittel at al. (1963b) irradiated in a fast reactor (EBR-I) specimens of cast hyperstoichiometric (U, Pu)C containing 17 to 28 wt% plutonium at power outputs roughly 750 W/cm length. However, since the fuel-surface temperatures were under 400°C, the centre temperatures were estimated to be under 500°C. Also, the exposures were under 0.1% burnup of the metal atoms. Apparent decreases in density around 1% were observed, but dimensional measurements were not sufficiently accurate to provide confirmation. Release of gaseous fission products did not exceed 1% of the theoretical yield. At the same time, 55% dense sintered specimens containing 20 wt% plutonium were irradiated under similar conditions. From changes in dimensions, densities were calculated to have decreased by as much as 1%, but confirmation by the immersion method was prevented by the large porosity. The most striking difference from the cast specimens was a fission-gas release of 12%, presumably due to the very high surface area of such porous material.

Kittel et al. irradiated some specimens of pure hyperstoichiometric PuC alongside those of mixed carbide. Apart from somewhat higher fuel temperatures, with the centres up to 650°C, the conditions were similar. Density changes around 1% were measured, but this time there were increases as well as decreases, while the dimensions remained unchanged. Thus, it is doubtful that apparent changes of 1% are significant under these circumstances. Release of gaseous fission products from the PuC was less than one percent.

Other specimens of cast (U, Pu)C, but hypostoichiometric and containing 10 wt% plutonium, have been irradiated at higher temperatures and to higher exposures by Bradbury et al. (1963): Two were at fuel temperatures around 1000°C for 0.6% burnup of the metal atoms, one was around 760°C for 1.25% burnup, and one was around 760°C for 1.65% burnup. No significant increases in volume were measured: However, since the specimens had been irradiated in tight-fitting stainless-steel sheaths, the fuel was restrained appreciably. Release of fission-product gases was under 2% for all four.

By comparing measured centre temperatures in nearly stoichiometric (U, Pu)C sinters, containing 20 wt% plutonium, with those in similar specimens of pure UC, Stahl and Strasser (1963) concluded that the thermal conductivity under irradiation of the mixed carbide was perhaps slightly lower than that of UC. An uncertain temperature drop between the fuel and the sheath prevented the determination of the magnitude of the conductivity, but the drop should have been the same for both materials. Post-irradiation examination of plutonium-bearing specimens of this composition that had been exposed to 3.3% burnup of the metal atoms at average centre temperatures calculated to be 1175°C showed volume changes up to 5% (Strasser et al., 1964). Although the centre temperatures reached a maximum of 1415°C, the fission-gas release did not exceed 1.5%, and there was no change in the grain size of the material.

Powder compacts (80% dense) of both (U, Pu)C, containing 20 wt% plutonium, and PuC were exposed to a maximum of 2.7% burnup of the metal atoms at power outputs up to 390 W/cm length and sheath temperatures up to 625°C (Neimark and Carlander, 1964a). The PuC powder, unlike the mixed carbide, had sintered almost to the sheath during operation. The release of gaseous fission products was about 30% from hypostoichiometric PuC, about 20% from hypostoichiometric mixed carbide, and only about 5% from hyperstoichiometric mixed carbide.

Additions up to 30% of PuC thus do not appear to change the properties of UC greatly: So far, the similarities in behaviour outweigh the differences. Specimens containing only a few percent plutonium, of potential application in thermal reactors, have not yet been investigated. The presence of plutonium, especially in the concentrations used for fast-reactor fuels, could conceivably make a significant difference (Runnalls, 1965). Since the free energy of formation is much less for the plutonium than the uranium monocarbide, several fission products that are unable to remove carbon from uranium may be able to do so from plutonium. Thus, metallic plutonium may be precipitated in specimens that have experienced a high burnup.

4.7 Other Fissionable Compounds, Including Uranium Nitride, Sulphide, and Silicides

Uranium Nitride and Carbonitrides

Uranium mononitride, like UC, has a face-centred cubic structure of the NaCl type. Its high uranium density (13.6 g/cm^3) and thermal conductivity close to that of UC make it attractive as a fuel, but the appreciable cross section of nitrogen for thermal neutrons has discouraged extensive development. Fabrication methods used for UC are also applicable to UN, but arc-melting must be performed in a nitrogen atmosphere to avoid substantial loss of nitrogen. Thermal decomposition, which becomes important over 2000°C, should not prevent UN being a satisfactory fuel.

Adam and Rogers (1961) included samples of UN when they investigated the effect of irradiation on the lattice parameter of UC (Sec. 4.3) and obtained very similar results for both materials. Irradiation of UN at a temperature of approximately 70°C caused an increase to an apparent saturation value of 1.6 parts in a thousand at an exposure of about 6×10^{16} fissions/cm^3. However, as for UC, there was an indication of an arrest in the increase about 7.5×10^{15} fissions/cm^3. Also, annealing UN irradiated to saturation damage revealed two recovery stages up to 640°C; one at about 150°C, and the other at about 550°C (Rogers and Adam, 1962b). From their observations, the authors suggested that two types of defect were present, with the second growing at a rate dependent on the concentration of the first. Although this study was restricted to exposures much lower than would be required of a reactor fuel, it showed that UN can suffer severe damage at low temperatures and suggested a similarity in behaviour to UC.

Bugl and Keller (1964) reported the successful irradiation of UN under conditions more appropriate to a reactor fuel. One of their specimens, whose densities ranged from 96 to 99% of that theoretically possible, achieved a burnup of 3.8% of the uranium (1.3×10^{21} fissions/cm³). At the start, the fuel temperatures for this specimen were estimated to be 615°C at the surface and 1260°C at the centre, but it was believed that they had come down to 255 and 575°C, respectively, by the end of the irradiation. For comparison with other fuels, these conditions correspond to power outputs of approximately 1900 to 600 W/cm length. The other specimens operated at considerably lower temperatures and were exposed to lower burnup.

Afterwards, some of the UN pellets were observed to be cracked, probably by thermal stresses but possibly by attack from the NaK in which they were immersed. In the central region of the hottest specimen, grain growth had occurred; porosity had diminished, and the remaining pores were rounded and situated on grain boundaries. In the same specimen, the fuel volume had increased by 3.5 to 6.4% — a moderate change for the relatively high burnup. For all specimens, gas release from the fuel was less than 1% of the krypton and xenon, but around 2.5% of the hydrogen calculated to have been produced by the $^{14}N(n, p)^{14}C$ reaction. Hardness increases of about 33% were measured on all specimens tested, regardless of burnup or irradiation temperature. Although little more than exploratory, these irradiations have confirmed the similarity to UC in behaviour and demonstrated no unforeseen problems.

Substituting carbon for nitrogen in UN gives compounds,

termed carbonitrides, that possess the same crystal structure as both UN and UC. Biddle (1964) lightly irradiated, at low temperatures, powders ranging in composition from $UN_{0.93}C_{0.07}$ to $UN_{0.1}C_{0.9}$, then measured the xenon evolved during post-irradiation annealing at temperatures from 800 to 1600°C. He also studied pure UN. As the composition varied at any given temperature, the diffusion coefficients calculated from these releases passed through a shallow minimum. However, the irradiation exposure of these samples was relatively high (3×10^{16} fissions/cm^3), and the coefficient at 1400°C for UC reported by Biddle was lower than corresponding values by most other investigators (Fig. 4–6). There is, therefore, a suspicion that internal trapping of the gas (Secs. 3.6 and 4.5) occurred in his samples. In experiments that measured the gas release during irradiation, Melehan and Gates (1964) obtained higher values than Biddle for the coefficient in pure UN, indistinguishable from those they obtained in UC (Fig. 4–6).

Uranium Sulphide

Like UC and UN, US has a cubic structure of the NaCl type and a relatively high thermal conductivity. Similar fabrication methods are suitable for all three. Although sulphur has a capture cross section for thermal neutrons only about one quarter that of nitrogen, the value, about 0.5 b, is still high enough to discourage the use of US in thermal reactors. Furthermore, the uranium density in US is no better than that in UO_2. These disadvantages partly explain the scarcity of irradiation experience with the material.

A series of six specimens of sintered US were irradiated by Neimark and Carlander (1964b). With a density 90% of that theoretically possible, exposure up to 8.2×10^{20} fissions/cm^3 (3.3% burnup of the uranium) caused volume increases $< 0.5\%$ and negligible fission-gas releases: The power outputs ranged from 400 to 800 W/cm length and the sheath temperatures from 400 to 800°C. Metallography revealed no significant change in the structure of these specimens. Another specimen, originally 80% dense, was bulged at the top, had developed internal porosity, and had released 20% of its fission gas, although nominally irradiated under similar conditions. However, the presence of UOS and UO_2 in the irradiated material, and the reaction of a tantalum spring with the fuel, suggested that this last result may not truly represent the behaviour of US under these irradiation conditions.

Hardness increases on irradiation were greater in US than in UN or UC. An increase of 75% resulted from irradiation at 600°C to 7×10^{15} fissions/cm^3, and a further 22%, from continuing the exposure to 1.5×10^{16} fissions/cm^3 (Pashos et al., 1964). The final material was reported to have had a melting point 50°C lower than that of unirradiated US (cf., Sec. 3.3 for UO_2).

Uranium Silicides

In the uranium-silicon system, the intermetallic compound with the greatest uranium content is the epsilon phase, U_3Si. This compound, with a body-centred tetragonal structure, has very narrow compositional limits and forms from a peritectoid reaction at 930°C. The combination in U_3Si of high uranium density, low parasitic absorption for thermal neutrons, and good resistance to aqueous corrosion naturally resulted in its potential as a fuel being investigated. The material is commonly prepared by melting and casting. It is sufficiently ductile to be extruded and may be coextruded in a Zircaloy sheath in a single operation.

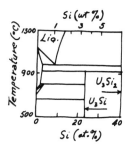

The most systematic study of the irradiation behaviour of U_3Si is still one by Bleiberg and Jones (1958). At temperatures under 250°C, they irradiated bare specimens immersed in NaK. A density decrease of 3.7 to 3.8% was measured for all exposures, even though these ranged from 0.01 to 0.1% of all atoms (5×10^{18} to 5×10^{19} fissions/cm^3). At the same time, hardnesses increased 40 to 100% with less variation in the final values than those before irradiation. Increases in the electrical resistivity were also

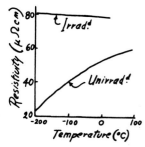

Temperature (°C)

independent of the exposure. The value for the irradiated material was about $80\mu\Omega$cm, virtually constant with temperature in the range -200 to $+20°C$, to be compared with 23 and $56\mu\Omega$cm, respectively, for unirradiated material at these temperatures. Metallography of irradiated samples revealed no grain boundaries, while no diffraction lines were exhibited in x-ray analysis.

If irradiation had caused decomposition of the U_3Si, alpha uranium and delta U_3Si_2 should have been detected by metallography and x-ray analysis. Since neither of these products was observed in a specimen exposed to 0.028% burnup of all atoms, Bleiberg and Jones suggested that the U_3Si structure was being disordered, even though thermal disordering of this compound has never been reported. They pointed out the structure's similarity to that of Cu_3Au, which is the classic example of the order-disorder phenomenon, and calculated a density decrease on disordering very close to that observed. They also argued that all of the specimen should have been affected after an exposure lower than any they tested, on the basis of reasonable assumptions that the average fission event produced six spikes, each involving 10^5 atoms. The large increases in hardness and resistivity would be partly due to the disordered structure and partly to internal stresses resulting from the volume changes that occurred piecemeal. The absence of grain boundaries is a natural consequence of the explanation, but the absence of diffraction lines must be ascribed to a very small crystallite size in the irradiated material. The fact that light surface abrasion markedly reduced the diffraction lines in an unirradiated specimen supported the hypothesis.

Two other series of irradiations have yielded results for comparison with those of Bleiberg and Jones. Kittel and Smith (1960) irradiated both as-cast and extruded U_3Si in similar capsules. Their specimens were estimated to have had surface temperatures from 90 to $650°C$, and centre temperatures from 100 to $900°C$: Exposures ranged from 0.07 to 0.45% burnup of all atoms. Generally, the specimens were deformed little by the irradiation and only a few surface cracks were apparent. The extruded material lengthened slightly more than the as-cast, but it was still much more stable than most uranium metal. Volume increases computed from dimensional changes are shown in Fig. 4–8. The specimen with the highest exposure was also the one with the highest operating temperatures: It was found to have developed a very large central void that accounted for much of its density decrease, so it has been excluded from the Figure. Some of the specimens were irradiated in coextruded sheaths, open at the ends. These showed smaller increases in diameter but greater increases in length, which indicates that the restraint can direct the fuel's expansion, as observed for uranium metal.

Howe (1960) irradiated coextruded U_3Si in a pressurized water loop with estimated fuel temperatures of 275 and 650°C at surface and centre, respectively, and measured the dimensional changes at interim examinations. The computed increases in volume of these specimens are compared with the others in Fig. 4–8. At 0.04% burnup of all atoms, the increase was only 1%, but it was around 6% by 0.09% burnup. In an earlier experiment, Howe (1965) had measured the electrical resistivity of small diameter specimens irradiated under similar conditions to a maximum of 0.04% burnup of all atoms: The values of both the resistivity and the temperature coefficient of resistivity were close to those observed by Bleiberg and Jones. However, when the latter investigators examined samples of Howe's fuel irradiated to about 0.03% of all atoms, they observed grain boundaries and U_3Si diffraction lines that had not been detectable in their own material, irradiated at approximately the same exposure.

Comparison of Howe's results with those of Bleiberg and Jones indicates that external pressure and elevated temperatures may prevent the disordering, or at least delay it to higher exposures. If Howe's results are considered on their own, the rapid increase in volume with increasing exposure and the large magnitude of some of the increases suggest the possibility that swelling associated

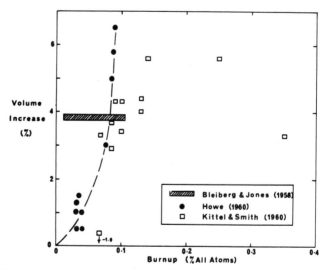

Fig. 4–8. The volume increase on irradiating U$_3$Si as a function of the burnup.

with gas-filled porosity might be responsible. However, such an explanation is difficult to reconcile with the observations by Kittel and Smith: One of their specimens irradiated to four times the exposure of Howe's, and at higher temperatures, had suffered less volume increase. Furthermore, even if the restraint offered by surface tension were neglected, the amount of fission-product gas generated by 0.09% burnup of all atoms would be insufficient to cause a volume increase of 6% against an external pressure of at least 100 atm. Thus, apart from the one high-temperature specimen irradiated to 0.45% burnup of all atoms by Kittel and Smith, disordering may still be largely responsible for the volume increases observed. Unfortunately, none of the investigations has included examination of the irradiated material by electron microscopy to look for microporosity. Potential users of U$_3$Si as a fuel have been deterred by the relatively large volume changes observed at low exposures and have not attempted higher exposures.

Concerning the resistance of U$_3$Si to aqueous corrosion even after irradiation, there is remarkable unanimity. Bleiberg and Jones, also Kittel and Smith, conducted

quantitative tests in static autoclaves as part of their examinations: Sheath ruptures gave Howe the opportunity to
make qualitative observations on the material's corrosion
resistance during irradiation. Specimens unsupported by
a sheath disintegrated on corrosion, but at temperatures
of 290°C, and under, the oxide film on the fragments was
adherent. The disintegration may, therefore, by largely
due to high internal stresses in a brittle material. At 315°C,
corrosive attack with hydrogen evolution was observed.
The absence of catastrophic attack in any of these tests
supports the belief that U_3Si does not decompose into
alpha uranium and delta U_3Si_2 under irradiation.

Even less is known about the irradiation behaviour of
other uranium-silicon compounds. Specimens of U_3Si_2
had increased in volume by roughly 7% and had released
2.5% of their fission-product gas after only 3.8×10^{19}
fissions/cm³ (0.4% burnup of all atoms) with an estimated
surface temperature of 670°C (Pashos et al., 1964).

Other Fissionable Compounds

In principle, any uranium compound may serve as a
fuel: Some of the more obvious possibilities are listed in
Table 4–1 along with a few of their relevant properties. A
similar list can be constructed for thorium compounds
(Griesenauer et al., 1964) and plutonium compounds
(Pardue et al., 1964). Additionally, a few atomic percent of
plutonium may be substituted for uranium in many compounds to provide even greater choice for potential fuels.
In practice, however, so much experience has been accumulated with UO_2 and UC that these two act as standards
by which the others are judged. For instance, a compound
that had a higher uranium density than that of UO_2,
while still possessing adequate corrosion resistance and
stability against irradiation damage without appreciable
parasitic neutron capture, would find ready acceptance
as a fuel for many reactors. Regardless of good irradiation
performance, many of the compounds of Table 4–1 have

Table 4-1
Some Uranium Compounds

Compound	Density (g/cm^3)	Uranium Density (g/cm^3)	Parasitic Cross Section (thermal) (b/U atom)	Possible Limiting Temperature (°C)
U_6Fe	17.7	17.2	0.4	815–peritectic
U_6Mn	17.8	17.2	2.2	726–peritectic
U_6Co	17.7	17.0	6.2	830–peritectic
U_6Ni	17.6	16.8	0.8	790–peritectic
U_2Ti	15.2	13.8	2.9	890–peritectic
U_5Sn_4	13.0	9.3	4.6	1500–melts
USi	10.4	9.2	0.2	1575–peritectic
UFe_2	13.2	9.1	5.2	1235–melts
UPb	14.5	7.7	0.2	1280–melts
USi_2	9	7.4	0.3	1700–melts
UAl_2	8.1	6.6	0.5	1590–melts
USi_3	8.3	6.1	0.4	1510–peritectic
UNi_5	11.3	5.1	23	1300–melts
UCu_5	10.6	4.5	19	1052–melts
UBe_{13}	4.4	2.8	0.1	2000–melts

never been tested in a reactor, since either UO_2 or UC would be preferable for a specific application. Although UZr_2 and zirconium-uranium hydride might be considered fissionable compounds, they are included with other zirconium-uranium alloys in Sec. 5.8.

5

Dilute and Dispersion Fuels

5.1 General Considerations

If the fissile material is dispersed in an inert diluent, the fission density and, hence, the fuel swelling can be kept to an acceptably low level while still achieving a large fractional burnup of the fissile atoms. If, in addition, the diluent is an isotropic material, the fuel is unlikely to be subject to irradiation growth with the resulting deformations, grain-boundary tearing, and enhanced creep rates found in uranium (Sec. 2). Niobium-rich niobium-uranium alloys are good examples of dilute fuels in which the uranium is dispersed on an atomic scale through the cubic niobium. More commonly, the fissile material, possibly as a compound, is dispersed as small particles in the diluent to form what is known as a dispersion fuel.

The principle of dispersion fuels is to concentrate the necessary fissile atoms in one phase dispersed in another, nonfissile phase. The dispersed phase can be a compound, such as UO_2, that is highly resistant to irradiation damage. The continuous phase must conduct the heat to the coolant, maintain structural and dimensional stability, contain fission products, and resist corrosion by the coolant. However, the fact that it no longer has to include uranium or plutonium opens up a much wider range of materials from which to select one with the optimum combination of properties for a particular application.

Obviously, the two phases must be chosen so that they do not react unfavourably during either fabrication or operation. There are, broadly, two methods of fabricating dispersion fuels: by synthesis from the component materials generally using powder-metallurgical techniques, and by precipitation of the fissile phase in a matrix that is nearly free of the fissile atoms, such as UAl_4 in nearly pure aluminum. In either instance, the ideal is to have each fuel particle individually contained in its own strong sheath.

The neutron economy of dispersion fuels is usually poor since the nonfissile phase not only absorbs neutrons but also occupies space that could otherwise be used for fissile or fertile material. These fuels, therefore, find application in reactors that cannot readily use uranium- or thorium-rich fuels. Where reactor physics dictates the use of highly enriched uranium, as for instance in the MTR-type reactor, dispersion of the fissile material in an inert diluent prevents excessive heat fluxes into the coolant or impracticably thin cross sections for the fuel. Also, in such highly enriched reactors, a burnup of well over 10 % of the uranium atoms is normally required, and, to achieve this without intolerable dimensional changes, considerable dilution of the uranium in an inert diluent may be necessary. Since a very small volume of the fissile phase is exposed to the coolant in the event of a localized sheath failure, dispersion fuels may be selected for reactors (such as the Army Package Power Reactor) that would be required to continue operating regardless of a few defects in the fuel sheaths.

For the fissile phase, uranium or plutonium, their oxides or carbides, or any of their other compounds considered in Chapter 4 are possibilities. Similarly, the nonfissile phase can be selected from any metal or ceramic that has a reasonably low cross section for neutron capture. (Since stainless steel with its high thermal cross section is commonly used, the interpretation of "reasonable" must be left to the reactor designer.) In practice, however, relatively few combinations have been considered and even fewer have been tested under irradiation.

Fission fragments recoiling from the surface of the fissile particles damage the surrounding matrix and eventually build up a concentration of impurities in it. The average range of a fission fragment is greater for a lighter fragment and for a matrix of lower density, but a value of 10μm is a fair approximation for most combinations. Ideally, each particle with its damaged layer should be separately enclosed in matrix material so that the fission products are retained and there is a continuous, relatively undamaged phase for structural strength and for heat conduction. Such a uniform dispersion is hard to achieve, but the development of coated particles (Sec. 5.5) has made it possible.

For a given volume fraction of fissile phase (usually determined by the limitations of fabrication methods), the volume of damaged matrix is proportional to the surface area of the particles. The fraction of the matrix subject to recoil damage can thus be minimized by the use of smooth, spherical particles as large as can be tolerated. Most fuels designed to take advantage of the dispersion principle have particles at least 100-μm diam. If, however, the matrix were much more resistant than the fissile phase to damage by the fission fragments, the size of the particles should be reduced to a minimum, and they should be acicular or tabular to increase their surface area. With a uniform distribution of spherical particles, the fraction of recoils escaping from the particles and the fraction of undamaged matrix can be calculated (see, for instance, the comprehensive review of the theory and practice of dispersion fuels by Weber, 1959) to yield the curves of Fig. 5–1.

5.2 UO₂-Steel Cermets

For thermal reactors in which neutron economy is not of prime concern, the good irradiation stability of UO_2 has been combined with stainless steel's good corrosion resistance in a variety of environments. UO_2-steel cermets are also a possible fuel for fast reactors using liquid metal as a coolant. Most of the UO_2-steel cermet fuels on which there is irradiation experience have been prepared by compaction of mixtures of UO_2 and steel powders. The

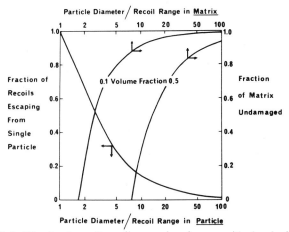

Fig. 5–1. The fraction of recoils escaping from an ideal spherical particle and the fraction of the matrix undamaged as functions of the ratio of the particle diameter to the recoil range in the particle and matrix, respectively. Derived from calculations by Weber (1959).

normal techniques of powder metallurgy permit up to 30 wt% (25 vol%) UO_2 to be incorporated without too much difficulty, while, with greater care, over 55 wt% (50 vol%) is feasible.

Many of the experimental irradiations of UO_2-stainless steel have been reviewed by Keller (1961). He exhibited the data by plotting each specimen on a graph of burnup against surface temperature: Figure 5–2 is a more recent version due to Kittel et al. (1964). For a typical specimen, containing 20 vol% UO_2, 10^{21} fissions/cm^3 is equivalent to 21.5% burnup of the uranium. On such a display, all the specimens that had failed by cracking or bulging during irradiation lay in the region of high burnup and high temperature. Therefore, Keller concluded that for each operating temperature there was a limiting burnup beyond which the *average* specimen became increasingly prone to failure.

Metallographic examination of these cermets suggests that the typical specimen could be greatly improved. The actual cross sections bear little resemblance to the ideal uniform dispersion of large spherical particles. In practice, the UO_2 particles are often angular and may be crushed

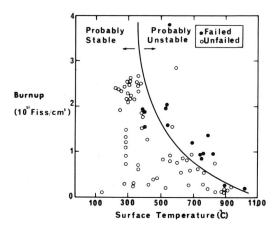

Fig. 5–2. Failure limits in dispersion fuels of UO_2 in stainless steel. (After Kittel et al. (1964), courtesy United Nations.)

during fabrication, so that large volumes of the matrix are subject to damage by fission-product recoil. Rolling, a process commonly used in fabrication, also tends to elongate the individual particles and can associate groups of particles in planar arrays. This defect, called "stringering", results in planes of weakness within the cermet. At high exposure, swelling of the UO_2, or release of the gaseous fission products from it, may cause bulging ("blistering") of the fuel at these locations (Fig. 5–4 in Sec. 5.3 and Fig. 5–9 in Sec. 5.6).

Goslee (1963) has attributed the progressively increasing burnup attainable from UO_2-stainless steel fuel to improvements in the fabrication processes: For instance, at temperatures around 400°C, early material was reliable up to 30% burnup of the uranium, but replacing the angular UO_2 particles with spherical ones and reducing the amount of cold rolling raised the limit to at least 60%. An even higher burnup may be possible with spherical particles individually coated with stainless steel before compaction, since such a process maintains the separation of the particles and largely eliminates their breakup and stringering.

The burnup attainable is also dependent on the percentage of fissile phase in the fuel. A specimen containing 26 wt% (21 vol%) UO_2 survived a burnup of 70% of the uranium at fuel temperatures about 550°C without failure (Richt and Schaffer, 1963). The density had decreased by 6%, due mainly to an increase in plate thickness. Metallography showed that the UO_2 particles had swelled sufficiently to account for the density change and that a few small cracks had developed in the matrix between adjacent particles. For comparison, another specimen containing 40 wt% (33.5 vol %) UO_2 had large cracks and was blistered by a similar exposure. Since the neck of matrix material between particles constitutes a point of weakness at which failures initiate, greater separation of the particles by incorporating less fissile phase delays the onset of failure by cracking. The shape and distribution of these UO_2 particles was not ideal, so still higher burnup should be attainable.

Frost et al. (1964) tested cermets containing 35 to 55 wt% (30 to 50 vol%) UO_2 or $(U, Pu)O_2$ in stainless steel. In corporating such a high concentration of fissile phase, the fabricators took special care to ensure a uniform distribution of spherical granules, several hundred microns in diameter. As a result, these specimens survived, without failure, irradiations at surface temperatures of 625°C to exposures over 10% of the heavy atoms.

A comprehensive metallographic study of the effects of high burnup on UO_2-steel dispersion fuel was reported by Barney and Wemple (1958), before there was comparable information available on bulk UO_2. Their specimens, which consisted of 20 wt% (15 vol%) of fully enriched UO_2 in stainless steel, were mostly irradiated with surface temperatures of 600°C and centre temperatures up to 780°C. These did not fail or show serious distortion, even at a burnup of 49% of the original uranium. By a few percent burnup, cracks in the UO_2 particles had healed while initial porosity within and around the particles had filled up: By 11% burnup a uniform distribution of fresh pores had become apparent, and by 49% burnup the pores had enlarged so that some were over 10 μm in diameter. One element, whose temperature was thought to have exceeded 1000°C during irradiation to 12% burnup, ruptured its sheath and had a crack to the centre of the fuel. The cross-sectional area of the fuel had increased by over 20%, so the UO_2 particles must have increased in volume by over 100%, a deduction consistent with the large amount of porosity seen in them.

Both lower temperatures and increased external restraint reduced the swelling of the UO_2 particles. Particles still confined within the steel matrix in the ruptured element that exceeded 1000°C showed greater porosity than those exposed to the same burnup at 600 to 780°C. Also, post-irradiation annealing at 900°C for 200 h of a specimen previously irradiated at the lower temperatures to 8.5% burnup caused an increase in the porosity of the UO_2. Particles exposed by the crack in the ruptured element, and therefore no longer restrained by the steel matrix, were much more porous than those still confined: There was no significant difference in temperature, since the exposed particles were cooled by NaK on the side that was not in contact with the matrix.

The shape and distribution of the large pores (over 10-μm diam) in some particles near the crack suggested that the UO_2 was extruding plastically out of the confining matrix (Fig. 5–3). The same behaviour was also seen in a specimen exposed to 20% burnup at 200°C. To explain other results (Sec. 3.5), irradiation-induced plasticity of UO_2 has been invoked at relatively low temperatures. However, in the present context, it is worth remembering the gross oversimplification involved in describing such highly irradiated material as "UO_2". By 20% burnup, there is one fission-product atom for every two uranium atoms, so that the fissile phase is a hodgepodge whose mechanical properties in the absence of irradiation are completely unknown.

Etching revealed a damaged layer, about 4-μm thick, in the steel matrix around each particle, presumably due to fission products recoiling from the UO_2. Metallography of cermets that had been deformed after irradiation showed

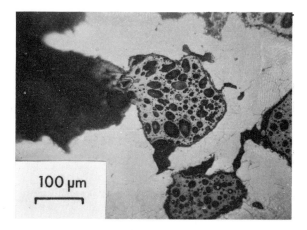

Fig. 5–3. Cermet of UO_2 in stainless steel irradiated to 11 % burnup of the uranium at an estimated temperature over 1000°C. The fuel fractured during irradiation and the shape of the pores suggests the UO_2 particle is extruding into the NaK coolant. (After Barney & Wemple (1958).)

that the layer was brittle. Even without any post-irradiation deformation some of the more highly irradiated specimens exhibited cracks in the steel, initiated at points in the damaged layer where a grain boundary in the steel met the UO_2 surface. In one batch of specimens, exposed to 20 % burnup, the etch attacked regions in the matrix around particles, but no uniform layer of damage was seen in this instance. The UO_2 particles in this batch were unusually small, about 30-μm diam, and the matrix unusual in being Type-347 stainless steel with a high silicon content of 1 wt %: The etching solution became enriched in silicon but not, to any appreciable extent, in fission products. The authors supposed that recoil of fission products left the UO_2 enriched in oxygen and that the effect would be large for these particular particles with their relatively high surface-to-volume ratio. Furthermore, silicon appears the element best able to compete for excess oxygen and probably diffuses in from the surrounding matrix, while oxygen diffuses out of the particles.

In another irradiation, Paprocki et al. (1961) observed a discontinuous grey phase, about 8-μm thick, surrounding UO_2 particles with a burnup of 45% of the uranium. The authors suggested that it might be a higher uranium oxide. However, the steel used for the matrix was a Type-347 containing niobium, and this may have reacted with the UO_2 to give the low melting-point eutectic in the $UO_2–Nb_2O_5$ system. Small, shiny, white particles were also seen distributed throughout the UO_2. These were similar to others in bulk UO_2 (Sec. 3.8) and were believed to be segregated fission products. The fact that the white particles have not been reported in several other UO_2 dispersion fuels irradiated to a high burnup can probably be attributed to differences in the metallographic preparation of the samples. Otherwise, the specimens of Paprocki et al. may have been operating at a higher temperature than they estimated — 370°C at the fuel surface.

Some further observations on such fuels have been described by Lambert (1963), who worked with 18.7 wt% (14.2 vol%) UO_2 in pure iron rather than stainless steel. His rod-shaped fuel, which contained 7% porosity in the as-fabricated condition, was irradiated at temperatures between 650 and 750°C to exposures ranging from 10 to 14% burnup of the uranium. Although the fuel was dimensionally stable during irradiation, subsequent annealing at temperatures over 900°C decreased the density, with a 4% decrease after 48 h at 1175°C. Optical and electron microscopy showed a uniform distribution of pores about 1-μm diam in the UO_2 particles of the as-irradiated fuel. Interfacial gaps that had existed in the unirradiated material were no longer detectable: This suggested that the UO_2 had swelled to fill the original porosity. Around each particle, a damaged layer in the matrix, about 5-μm thick, was revealed by its different etching characteristics and its increased hardness. Electron microscopy showed that very fine pores were just beginning to precipitate in this layer. After annealing, some of those pores and the ones in the UO_2 had grown to several microns diameter: Together, they accounted for much of the observed decrease in density. At isolated places, pores in the interface had linked up to cause a gap between the fissile and matrix phases of the irradiated and annealed material.

Observations on these cermets can be regarded as a further source of information on the irradiation behaviour of UO_2. Thus, swelling of the UO_2 is small for exposures up to 10 at.% burnup of the uranium at 750°C, where there is considerable restraint from the matrix, while increasing the burnup or temperature or decreasing the restraint results in appreciable swelling.

5.3 Uranium Oxides in Aluminum

For application in most low-temperature coolants, aluminum can replace steel as the matrix material, thereby greatly improving the neutron economy. In UO_2-aluminum dispersion fuels, swelling of the UO_2 occurs in the same manner as it does in UO_2-steel. In plate elements containing 32 wt% (11 vol%) of fully enriched UO_2 irradiated to 6×10^{20} fissions/cm^3 (25% burnup of the uranium) at fuel central temperatures about 150°C, a fine distribution of porosity was seen in the UO_2 by optical microscopy: In others containing 41 wt% (16 vol%) irradiated to 1.25×10^{21} fissions/cm^3 (33% burnup) at220°C, the pores were about 10-μm diam (Graber et al., 1964).

In the aluminum matrix, the UO_2 particles had contracted markedly and become nearly spherical, while etching revealed that the region between the remaining UO_2 and the aluminum was composed of two distinct zones. Although the thickness of the combined zones is comparable to the recoil range for fission fragments, it seems more likely that the zones represent products of the UO_2/aluminum reaction. In the absence of irradiation, UO_2 and aluminum react appreciably at 600°C with thy formation of Al_2O_3, UAl_4, and sometimes UAl_3: Fabrication methods for UO_2-aluminum fuels are considerable restricted by the desire to avoid this reaction. The two reaction zones in the irradiated material have not been positively identified, but the outer one that etches rapidly may be a uranium-aluminum compound, while the harder and more resistant inner one may by Al_2O_3. The latter, if present, would be expected to be amorphous (Sec. 3.12). As a result of the reaction, the volume fraction of the matrix phase decreases greatly. In the specimens irradiated by Graber et al., the initial UO_2 contents were 11 and 16 vol%, but, after irradiation, the reacted particles occupied about 50 and 65 vol%, respectively.

Although the more dilute fuel survived its irradiation without any gross damage, all plates made from the 41 wt% material, which was irradiated to a greater burnup at a higher temperature, exhibited one or more blisters. Cross sections showed that individual particles of UO_2, in reacting and swelling, had initiated cracks in the surrounding matrix (Fig. 5–4). Where there was stringering, a large crack ran through many adjacent particles. Bulging of the crack to form a blister was presumably caused by the pressure of fission-product gases released from the UO_2, since the appropriate amount of these gases was obtained from some of the blisters on subsequent puncturing. Thus, an improved fabrication route that removed stringering should increase the burnup obtainable from UO_2-aluminum, but the attainable volume fraction of UO_2 will probably remain lower than for UO_2-steel because of the reaction zone.

Thermodynamically, U_3O_8 also is unstable in the presence of aluminum. However, their first reaction is to form UO_2 and Al_2O_3, with the latter apparently providing a barrier that slows down further reaction. In practice, therefore, U_3O_8-aluminum can be fabricated and operated at higher temperatures than can UO_2-aluminum. Graber et al. included some U_3O_8-aluminum plate elements in their tests: 35 wt% (15 vol%) of fully enriched U_3O_8 in aluminum exposed at a fuel central temperature of 180°C to 7.5×10^{20} fissions/cm^3 (29% burnup of the uranium), and 44 wt% (21 vol%) at 205°C to 1.1×10^{21} fissions/cm^3 (33% burnup). Once again, as a result of its reaction with the particles, the volume fraction of the matrix decreased

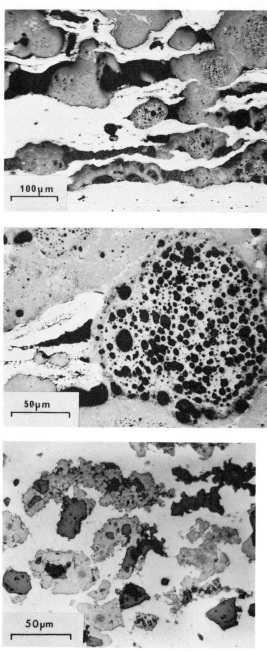

(a)

(b)

(c)

Fig. 5-4. Cermets of Al–41wt%UO$_2$ — initially 16vol%UO$_2$.
(a) Cracks in the aluminum matrix formed by swelling and reaction
    of UO$_2$ particles.
(b) Porosity in UO$_2$ particle.
(c) Unirradiated material heat treated at 590°C.

(After Graber et al. (1964).)

to 55 and 30%, respectively. A reaction zone surrounded each U_3O_8 particle, which had diminished in size and contained distributed porosity. In these specimens, however, the combined U_3O_8 and reaction products had separated from the matrix, leaving large voids at the interface. The pressure of fission-product gases released into the voids presumably ruptured the necks of aluminum between stringered particles to produce the blisters seen in all the elements containing 44 wt% U_3O_8.

Reinke (1963) irradiated specimens of aluminum-39 wt% (19 vol%) U_3O_8 at temperatures under 100°C to various exposures up to 10^{20} fissions/cm³ (4% burnup of the uranium). By exposures of 2×10^{19} fissions/cm³ (slightly under one percent burnup), the originally porous U_3O_8 particles had sintered: This concentrated the pores into a few large voids (Fig. 5–5). There was a thin layer (about 2-μm thick) of reaction product around each particle: This layer was resistant to HF etchant and may have been Al_2O_3. None of the specimens had suffered any gross damage during irradiation, and all but one had increased in density by small amounts up to 0.5%. Subsequent annealing at 550°C caused a density decrease of about 5%, but unirradiated controls decreased in density by over 3%. Thus, part of the decrease may have been due to the U_3O_8-aluminum reaction or even to release of hydrogen impurities into the porosity.

The UO_2- and U_3O_8-aluminum cermets are therefore susceptible to the same blistering type of failures that occur with UO_2-steel. In addition, reaction of the aluminum with the uranium oxide can produce a further volume increase. The reaction of the components, together with the mechanical and corrosion properties of aluminum, all restrict operation of the aluminum-based cermets to relatively low temperatures.

5.4 Other Cermet Fuels

The investigation of UO_2 dispersions in refractory metals has been stimulated by the desire to extract the heat

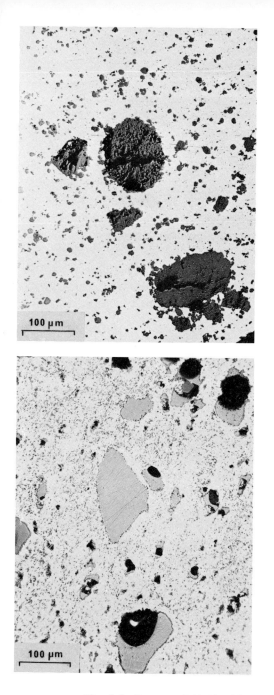

(a)

(b)

Fig. 5–5. Cermets of Al–39wt%U_3O_8.
(a) Unirradiated.
(b) Irradiated.

(After Reinke (1963).)

at higher temperatures and to develop nuclear-heated thermionic emitters for direct conversion. Keller et al. (1963) irradiated rods of Mo-81wt%UO_2 and Nb-82wt%-UO_2 (both 80 vol% UO_2) to about 4% burnup of the uranium. The UO_2 consisted of porous, spherical particles around 100-μm diam. Despite the high UO_2 contents, special fabrication techniques had resulted in the metals forming a continuous phase. The molybdenum-UO_2 was prepared by powder-pressing the constituents, with the result that adjacent fissile particles were in contact in the final product: Vapour deposition of niobium on the UO_2 particles before pressing produced a metal layer, of minimum thickness about 10 μm, between these particles.

At the start of the irradiation, fuel-surface temperatures for the various specimens lay in the range 925 to 1200°C, with centre temperatures estimated to be from 1400 to 1975°C, but the loss of a NaK heat-transfer bond during the irradiation caused most of these temperatures to be exceeded. The use of strong sheaths and generous end clearances prevented much being learned of the fuel's behaviour, since post-irradiation dimensions were taken on only the unopened capsules. Metallography showed the outer regions of the fuel to be largely unchanged, while the central region of some rods was substantially altered. At the unknown high temperatures both the molybdenum and niobium had become dispersed in a continuous phase of UO_2, which exhibited grain growth.

Rods of tungsten containing 50 wt% (67 vol%) UO_2 have been irradiated for about 3 h at temperatures over 2100°C without any obvious reaction occurring between the fissile and matrix phases (Flaherty and Panisko, 1964). Other specimens of W-50wt%UN (57 vol%) and Mo–72wt%UN (65 vol%) survived 21 days irradiation to an exposure roughly 2×10^{19} fissions/cm^3 at temperatures over 2000°C without apparent damage: However, the release of fission-product gases was appreciable — up to 60% (Flaherty et al., 1964).

5.5 Coated Particles

The high-temperature capabilities of ceramic fuels offer the opportunity of increased thermodynamic efficiency by raising coolant temperatures. Retention of active fission products within the fuel body is still necessary, yet the performance of the low-absorption metals normally used for sheathing is inadequate at temperatures around 1000°C. Simple replacement of the metal sheaths by tubes of graphite, beryllia, etc., while essentially retaining the conventional geometry, has been considered. Porous graphite, homogeneously impregnated with a fissile compound, is another possibility. Typically, the graphite is impregnated with uranium nitrate which is then fired to yield the carbide. Cubicciotti irradiated such specimens as early as 1952: His and subsequent experience, summarized by Bromley (1963), showed that the principal disadvantage of impregnated graphite was its failure to retain fission products.

Since 1960, however, development of coated particle fuels has demonstrated the value of a new approach. Small, nearly spherical particles of fuel are individually encased in impervio s ceramic coatings and then distributed in an inert matrix. Particle diameters of 100 to 500 μm are normal, with coating thicknesses of about 100 μm. The oxides or carbides of uranium and thorium have been used for fuel, while pyrolytic carbon, Al_2O_3, BeO, MgO, SiC, and ZrC are among the possible materials for coatings. Graphite and BeO are favoured as matrix materials. The final product is, literally, a dispersion fuel element, but the function of the inert coating and matrix is slightly different from that of the metallic matrix in the earlier dispersion fuels.

The single requirement of the coating is that it contain, or retain, the active fission products both gaseous and solid. To that end it must be compatible with the fuel and matrix materials and with the coolant, including any impurities in the latter. The coating must stop within its thickness any recoils from the fuel and must prevent diffusion of the fuel or fission products through its wall. Finally, it must remain intact despite thermal stresses, irradiation damage, and internal pressure due to released gases or fuel swelling. Perhaps surprisingly, the requirement can be satisfied for a burnup of about one quarter of the metal atoms at temperatures around 1000°C.

An upper limit to the operating temperature for certain fuel/coating combinations can be defined from known diffusion rates. If oxygen diffuses through the coating of UO_2 or carbide particles, the fuel is oxidized and the consequent expansion can cause failure. In an oxidizing environment, only Al_2O_3 and BeO appear suitable, and then only for temperatures below 1100°C. Uranium diffuses through pyrolytic carbon appreciably over 1600°C. In a thermal gradient at temperatures over 1700°C, UC_2 particles have migrated out of their pyrolytic carbon coatings by eating them away on the hotter side and depositing carbon between fuel and coating on the cooler side (Goeddel, 1962): Excess carbon in the UC_2 is reported to suppress the migration, whose mechanism has not yet been explained. Townley et al. (1964) have calculated that, for fission products with half-lives shorter than 12 days, diffusion coefficients of less than 10^{-13}cm^2/sec are required to prevent excessive release of activity. This criterion applied to experimentally determined coefficients for ^{133}Xe yielded a temperature limit of 1500°C for Al_2O_3, BeO, MgO, and columnar pyrolytic carbon and 1200°C for the laminar form of pyrolytic carbon. However, Dayton et al., (1964) have shown that fission products such as barium, iodine, silver, tellurium, cerium, and neodymium diffuse much more rapidly than xenon in pyrolytic carbon, which restricts the upper limit to below 1000°C in that material, if activity release is to be avoided: Al_2O_3 and BeO are much more effective in retaining these fission products.

Irradiation tests have confirmed that the release of rare gases through the coatings is negligible at temperatures around 1000°C. The continuous fractional release from uncoated particles of UC_2 at 815°C was about 10^{-2} of that being generated, while it was only about 10^{-7} from coated particles at 1300°C (Townley et al.). Other tests showed that where appreciable releases are observed, they are the consequence of mechanical failures of the coatings. The fraction of gas collected on post-irradiation puncture of capsules containing coated particles correlated well with the fraction of particles later found to have failed (Sayers et al., 1963). More directly, Harms (1962) showed that the number of bursts of activity indicated by continuous monitoring during post-irradiation annealing corresponded with the number of failed particles. Thus, before diffusional release need be considered, integrity of the coating has to be guaranteed.

Sayers et al. reported structural changes in carbide particles as a result of irradiation. Particles that were originally composed of UC_2 plus small amounts of UC, after irradiation to several percent burnup of the uranium atoms at 800 to 1300°C, appeared to consist of a carbon precipitate in a carbide matrix (Fig. 5–6b). This recalls observations on bulk $(UC + UC_2)$ by Freas et al. (Sec. 4.2). The wall thickness of the pyrolytic carbon coating was unchanged, indicating that no extra carbon had entered the fuel. Microhardness measurements suggested that the UC_2 had transformed to UC or U_2C_3, but x-ray analysis of similar specimens demonstrated that both UC_2 and UC were still present. Sintered particles densified during irradiation. The effect was greatest at 700 to 800°C: At higher irradiation temperatures, the carbide phase became almost fully dense, but porosity was apparent in the precipitate that was thought to be carbon. The dense fuel appeared to have flowed plastically into notches in the inner surface of the coating, which suggests irradiation-enhanced creep. The fuel seemed to be swelling appreciably, but normally no porosity could be detected in the carbide phase by optical microscopy.

In one series of specimens irradiated to nine percent burnup of the metal atoms at 800°C, all failed particles exhibited gross pores up to 25-μm diam (Fig. 5–6). However, since similar particles survived a burnup of 30% at 900°C without damage to the coating or obvious swelling, the coating failures are probably the cause of the swelling rather than its consequence (Bomar and Gray, 1964). In

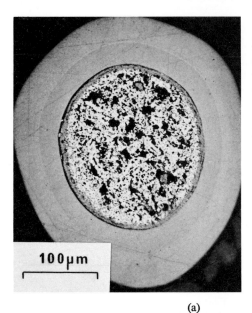

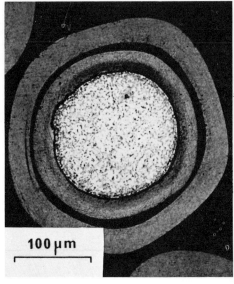

(a)   (b)

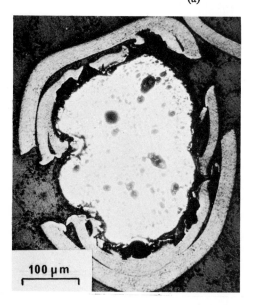

(c)

Fig. 5–6. UC₂ particles coated with pyrolytic carbon.
(a) Unirradiated particle.
(b) Irradiated, intact particle.
(c) Irradiated particle; porous fuel, ruptured coating.
    (After Sayers et al. (1963), courtesy Macmillan and Co. Ltd.)

their high burnup specimens, Bomar and Grey observed a reaction zone, presumably rich in fission products, just inside the coating. Sayers et al. sometimes observed a reaction zone, but of a different type, due apparently to interdiffusion of the fuel and coating materials: Where the reaction occurred, the sintered fuel did not densify during irradiation.

The same irradiations by Sayers et al. included specimens of mixed uranium and thorium dicarbide with a Th/U atomic ratio of 5:1 or 10:1. Porous sinters densified during irradiation at 1150 and 1375°C. Particles that had been prepared by melting were single crystals before irradiation, but had a striated microstructure. Recrystallization appeared to have occurred in those irradiated at 1400°C and over, but not at 1300°C. Even in unirradiated material, the equilibrium structures in this region are not well established. At the lower irradiation temperatures, cracks often formed along crystallographic planes in the carbide single crystal and propagated through the coating, but observations of similar effects during the coating process indicate that the cracking is not primarily due to irradiation. Specimens of mixed uranium and zirconium monocarbide, with a Zr/U atomic ratio of 5:1, were polycrystalline whether prepared by sintering or melting. Either type showed little effect of irradiation to nearly 3% burnup of the metal atoms in the range 1350 to 1750°C.

Townley et al. (1964) reviewed the behaviour of oxide-coated particles. One batch of Al_2O_3-coated UO_2 particles had a very low continuous fractional release for rare gases of under 10^{-7} during irradiation to 10% burnup of the uranium at 1025°C. This excellent performance was obtained with relatively thin coatings (42-μm) and high-density fuel (2% porosity). However, when more particles of the same batch were irradiated at 850°C, 1 or 2% had failed by 7% burnup of the uranium. Other experiments have shown that decreasing the fuel density or increasing the coating thickness diminishes the incidence of failures. Varying the specimen temperature during irradiation established that the particles were more prone to fail at temperatures below 500°C, than above. This observation makes it unlikely that internal pressure on the coating, due to fuel swelling and gas release, is alone responsible for the failures.

Elleman et al. (1963) have shown that fission fragments recoiling into the coating may cause sufficiently high integral stresses to rupture the coating. Discs of Al_2O_3,

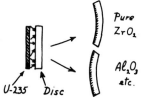

BeO, MgO, SiO_2, pure and stabilized ZrO_2, SiC, and pyrolytic carbon were bombarded on one face by about 10^{17} fission fragments/cm³ of surface layer at a temperature near 50°C. Interferometric measurements of the induced changes in curvature of the discs revealed that the damaged layer had contracted in pure ZrO_2 but had expanded in all the others. X-ray diffraction showed that the changes were not due to the damaged layer being rendered amorphous (cf., Al_2O_3–UO_2, Sec. 3.12). Subsequent annealing of some established that most of the damage disappeared from SiO_2 at 800°C and from Al_2O_3 and BeO at 1200°C: Annealing at 600°C removed the curvature from both pure and stabilized ZrO_2, but, at higher temperatures, the curvature reversed in both materials. In the pyrolytic carbon, deformations occurring at 400°C prevented measurements being made. Thus, when used as coatings, any of these materials except pure ZrO_2 would be in compression near the inner surface but in tension further out. The stresses would be greater at low temperatures where annealing is negligible.

The same tendency for failure to occur more readily at low temperatures has been observed with BeO and pyrolytic carbon coatings. Irradiations of UC_2 particles coated with pyrolytic carbon demonstrated that thermal cycling between 650 and 1100°C did not initially cause failures, but only after a few atomic percent burnup (Burian et al., 1963). Other carbon-coated particles, examined by Sayers et al. (1963), yielded more direct evidence of damage due to recoils. Irradiation had caused separation and shrinkage of an inner layer of coating, closely similar in thickness to the recoil range for fission fragments in carbon (Fig. 5–7). The strong response of this layer to polarized light suggested that the pyrolytic carbon had densified as a result of becoming ordered, or even graphitized. Characteristic cracks, termed "spearheads" by the authors, were present in the damaged layer: On a diametral plane, these were V-shaped in section, pointing outwards, while on a tangential plane their cross sections were X-shaped. Microradiography confirmed both the separation of the inner layer and the presence of the spearhead cracks. Taken in conjunction with the results by Elleman et al., these observations indicate that, while recoil damage expands pyrolytic carbon at low temperatures, large contractions occur above some temperature in the range 100 to 400°C.

In some particles, spearheads penetrated the entire thickness of the pyrolytic carbon coating (Fig. 5–7). In others, the coating was cracked radially in a brittle fashion: It appeared that internal pressure had caused a crack to initiate at a spearhead. In yet other particles, fine cracks in the coating that are neither radially oriented nor associated with a spearhead have been observed. To decrease failure rates, both thicker coatings and laminated coatings have been employed (Dayton et al., 1964). Particles with a 40-μm thick inner layer of porous carbon overlaid with a 70-μm thick coating of pyrolytic carbon survived to 5% burnup, even when thermally cycled up to 1000°C. Where an inner layer of laminar pyrolytic carbon was overlaid with columnar pyrolytic carbon, 26% burnup was achieved (13% at 815°C, then 13% at 925°C), with less than 0.5% of the particles cracking. Aluminum oxide has been successfully used as an inner layer on UO_2 with pyrolytic carbon as the coating. Microscopy shows that the laminated coatings help to limit the propagation of spearhead cracks.

Irradiations already performed prove that coated particles can survive intact over 20% burnup at temperatures as high as 1300°C (Fig. 5–8). However, further developments may be necessary to ensure full reliability under such stringent conditions. An interlayer of SiC in a graphite coating is effective in stopping the penetration of many of the solid fission products, but this barrier layer

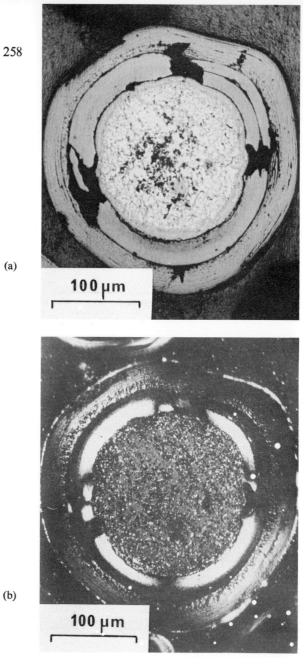

(a)

(b)

Fig. 5–7. Examples of damage observed in irradiated coated particles:
See also Fig. 5–6.

(a) Inner layer of pyrolytic carbon shrunk and separated, under
white light.

(b) As (a), but under polarized light.

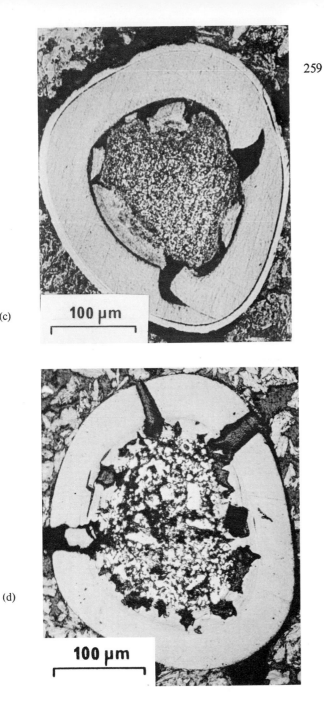

(c) Large spearhead cracks nearly penetrating the coating.
(d) Wide radial cracks through the coating.
    (After Sayers et al. (1963), courtesy Macmillan and Co. Ltd.)

260

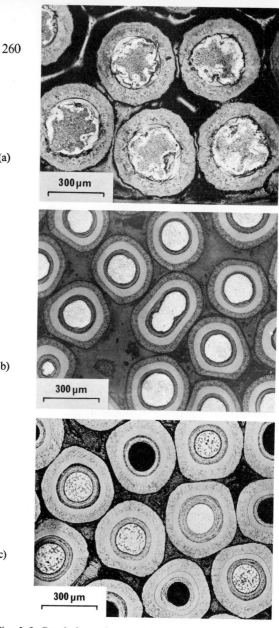

(a)

300 µm

(b)

300 µm

(c)

300 µm

Fig. 5–8. Pyrolytic carbon-coated particles successfully irradiated to high burnup.

(a) UC$_2$ after 26% burnup of the uranium at 815 to 925°C.

(b) UC$_2$ after 24% burnup of the uranium at 1370°C.

(c) UO$_2$ after 25% burnup of the uranium at 1600°C. (Some of the coatings are empty because particles fell out during preparation for microscopy.)

(Courtesy W. O. Harms & E. L. Long, Oak Ridge National Laboratory.)

has been seen to be cracked in some of the particles sectioned. Whether the cracks permit escape of the solid fission products remains to be established. Further improvements in the irradiation performance of coated particles may result from recent laboratory studies: Large variations in the density, structure, and anisotropy of pyrolytic graphite can be obtained by suitable control of the conditions for deposition.

5.6 Aluminum-Uranium and Magnesium-Uranium

Aluminum-uranium fuels normally contain less than 50 wt% (about 10 at.%) uranium, and consist of angular particles of UAl_4 dispersed in nearly pure aluminum. These alloys can be readily cast and extruded, while those containing up to 25 wt% uranium can be rolled and bonded to aluminum sheathing to form plate elements. The latter have found extensive application in experimental and test reactors of the MTR type, where the coolant is cold water. Aluminum-uranium fuels can also be prepared by powder compaction, when the fissile particles may be a mixture of UAl_4 and UAl_3.

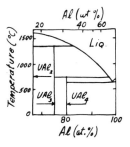

When irradiated at temperatures under 100°C, alloys of about 20 wt% uranium suffer no appreciable dimensional changes for exposures as high as 70% burnup of the uranium. However, some other physical properties of the fuel are seriously affected: Thermal and electrical conductivities can be reduced to about half their initial values, the hardness and tensile strength can increase severalfold, while the uniform elongation decreases from several tens percent to less than 5%. Much of the damage does not anneal until 300 to 600°C, although fast neutron damage in pure aluminum anneals at −60°C. Metallography revealed very little structural change in the as-irradiated alloys, even at 70% burnup (Gibson and Francis, 1962).

Reinke (1963) annealed specimens of Al–17.3wt%U (Al–2.3at.%U) previously irradiated below 100°C. Where the exposure was 1.4×10^{20} fissions/cm³ (10% burnup of the uranium) no swelling was detected for annealing temperatures up to 550°C, but, for 5.9×10^{20} fissions/cm^3, swelling was rapid at the same temperature. Metallography showed that voids along lines of UAl$_4$ particles probably joined up to form a macroscopic blister (Fig. 5–9). If such a mechanism is responsible for the gross swelling, the critical combination of exposure and temperature may be greatly affected by the fabrication route, with a uniform dispersion of UAl$_4$ expected to give the best performance. The 550°C anneal had little effect on the morphology of the angular UAl$_4$ particles in the material exposed to 1.4×10^{20} fissions/cm^3, but it caused major disruption of the particles at the higher exposure.

Annealing Al–20wt%U alloys that had been lightly irradiated, Reynolds (1958) found that the release of the fission-product gases was negligible below the eutectic temperature of 640°C. Above, however, a large fraction being released: After an initial burst, the amount collected increased as the square root of the annealing time. The burst from a specimen annealed first below then above the eutectic temperature consisted of two stages, the second being equal in size to the single burst from a specimen annealed only at the higher temperature. Reynolds also obtained measurable gas release from separated UAl$_4$ powder held at 600°C. These observations, together with the metallographic ones, suggest that the fission-product gases are largely retained within the bigger UAl$_4$ particles during irradiation at temperatures around 100°C. In the range 500 to 600°C, they are sufficiently mobile to escape from the particles but appear unable to enter the surrounding aluminum matrix. Failure by blistering would then occur when the pressure in adjacent gas pockets caused rupture of the intervening matrix.

Since, under irradiation, rare gases can pass from a gas phase into a solid (Sec. 3.6), the annealing results may not be directly applicable to reactor fuel. However, Gavin and Crothers (1960) reported a similar failure of a fuel plate of Al–2wt%Ni–17.5wt%U irradiated to about 50% burnup of the uranium when the estimated temperature

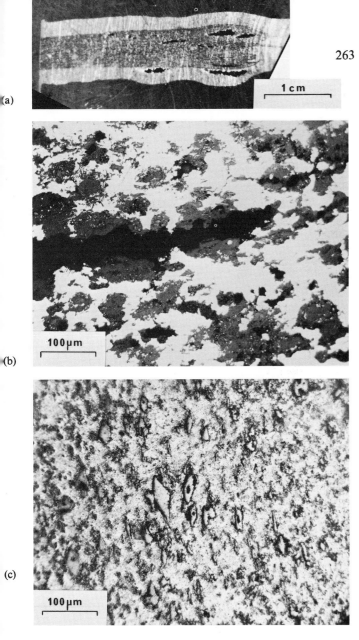

Fig. 5–9. Al–17.3wt%U plates irradiated below 100°C then annealed at 550°C.

(a) Cross section of plate irradiated to 5.9×10^{220} fissions/cm^3 showing blisters.

(b) Blister at higher magnification.

(c) Cross section of plate irradiated to 1.4×10^{20} fissions/cm^3 showing UAl$_4$ particles.

(After Reinke (1963).)

of the fuel in the reactor was near 550°C. The apparent agreement of the temperature limit under different conditions could be fortuitous and requires confirmation on more samples.

Runnalls and Boucher (1965) have suggested that the good resistance to swelling of aluminum-uranium alloys is due to fission products being accommodated in uranium vacancies of the UAl_4 structure. Since roughly 10% of the uranium sites are vacant in the unirradiated compound, there may be room for fission products within its lattice up to a burnup of 10% of the uranium. Runnalls and Boucher have shown that below approximately 646°C the vacancies can be ordered and have suggested that the gas releases from UAl_4 observed by Reynolds may have been associated with ordering or disordering of the structure. Further gas-release measurements, in the range 600 to 646°C, from well-characterized UAl_4 would be required to confirm this interpretation, which predicts good swelling resistance for aluminum-uranium alloys up to 10% burnup even at elevated temperatures.

In the magnesium-uranium system, there is virtually no solubility of either element in the other, and no intermetallic compounds. Thus, uranium particles can be dispersed in a magnesium matrix without fear of interaction. In 1952, H. H. Hirsch suggested the use of magnesium-uranium fuel to overcome the deformations then being experienced on irradiating uranium rods. Freshley and Last (1956) reported the results of irradiating compacts of 50 vol% of uranium chips, of roughly 1-mm size, that had been impregnated with magnesium. These were exposed to a burnup from 0.1 to 2.0% of the uranium at estimated temperatures of 320°C at the fuel surface and 390°C at the centre. Even at the highest burnup, there was no distortion or expansion measured, while a uranium rod would have been badly damaged under the same conditions. The same specimens showed a small increase in strength and considerable loss of ductility following irradiation: Most of the effect had already occurred after the lowest burnup. Using uranium shot instead of chips, the same authors achieved a uranium content of 64 vol%: Such a specimen sustained a burnup of 0.6% of the uranium at estimated surface and centre temperatures of 330 and 500°C, respectively, without failure. In theory, the individual particles of these fuels are free to deform under irradiation, but plastic flow of the magnesium prevents any gross distortion of the fuel as a whole: Unfortunately, metallography to establish this point was not performed on the irradiated specimens.

5.7 Thorium-Uranium

Unlike other potential diluents, thorium has an appreciable absorption cross section for neutrons. However, the capture results in the production of the fissile isotope ^{233}U, so the neutrons are not wasted. Up to 1000°C, about 1% of uranium goes into solid solution, and further amounts form a uranium precipitate predominantly at grain boundaries. Compositions up to approximately 20% uranium can be regarded as dispersions of uranium in a face-centered cubic matrix, while, at higher uranium contents, the uranium may form a continuous network.

Experiments by Kittel et al. (1963a) have demonstrated that excellent irradiation stability is possible in these alloys. Samples of eight different compositions containing up to 31 wt% uranium were irradiated to several percent burnup of all atoms with centre temperatures up to 1000°C. Below 500°C, the volume increase per 1% burnup was 1 to 2% (Fig. 5-10). At higher temperatures the swelling was more variable and increased, but only slightly. Even at 1000°C (and possibly higher) the volume increase was only 8.5% per atomic percent burnup. The specimens containing 25 and 31 wt% uranium exhibited surface roughening, distortion, and sometimes cracks: Those containing 20 wt%, and less, uranium showed no such damage. Although the high-uranium specimens were generally subjected to more severe conditions than the low-uranium ones, whose centre temperatures did not exceed 800°C and whose burnup was under 5 at.%, the

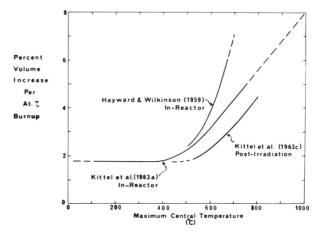

Fig. 5–10. The volume increase per unit burnup of thorium-uranium alloys as a function of the irradiation temperature.

overlap was sufficient to establish the superior behaviour of the latter. However, the composition had no apparent effect on the volume changes of these alloys.

Kittel et al. (1963c) annealed samples containing 10 to 25 wt% uranium previously irradiated to at least 4% burnup of all atoms at 640°C or lower. Measuring the density after step anneals, they found volume increases of about 4% per 1% burnup at 775°C: This is in good agreement with the in-reactor results.

Optical microscopy revealed little change in the structure due to irradiation. Generally, the uranium precipitate tended to agglomerate more at the grain boundaries, and cracks were seen in the Th–31wt%U alloy. Unless the specimen was cracked, thorium-uranium alloys released < 0.5% of the fission-product gases generated (Gates et al., 1959). As in other metals, the irradiation causes increases in hardness and tensile strength, with accompanying decrease in ductility. Annealing at 750°C produces partial recovery.

Kittel et al. (1963a) attributed the good stability of these fine dispersions to the fact that most of the damage caused by recoiling fission fragments occurs within the thorium matrix. They calculated an escape probability of about 90%

for recoils from uranium particles 1-µm thick. The thorium, being face-centred cubic, is not susceptible to growth and associated phenomena seen in anisotropic uranium, so it should be better able to withstand the damage. Another possibility is that the fine dispersion of uranium particles helps to anchor the gas-filled pores (Sec. 2.5). Also, the higher melting point of thorium (1755°C) may reflect lower diffusion rates at any given temperature, with the consequent possibility of slower rates of swelling. Kittel et al. compared their results with earlier ones on Th–11wt%U that exhibited an increase in swelling at lower temperatures (Fig. 5–10): They suggested that the thicker uranium particles (5- to 10-µm) observed in the specimens reported by Hayward and Wilkinson (1959), by retaining more of the recoil damage, were responsible for the poorer performance. If this explanation is valid, the general principle of dispersion fuels, that the fission damage should be confined to the fissile phase as far as possible, would not be applicable to thorium-uranium alloys. Surface roughening and distortions were associated with compositions where the uranium formed a continuous network, while negligible deformations occurred on irradiating specimens with the uranium particles isolated.

5.8 Zirconium-Uranium and Zirconium-Uranium Hydride

Because of the good corrosion resistance of zirconium alloys, zirconium-uranium is an obvious possibility for application in a hot-water environment. In the composition range 5 to 25 wt% uranium that has been most thoroughly investigated, the intermetallic delta-phase UZr$_2$ is normally dispersed in an epsilon-phase matrix of nearly pure zirconium. Thus, the fuel can be bonded to zirconium-alloy sheaths. The $(\delta + \varepsilon)$ transforms to $(\gamma_2 + \varepsilon)$ above 595°C, but the latter can be retained on quenching. Compositions entirely within the delta phase have also been tested as fuels.

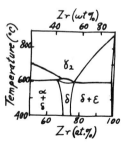

In an extensive series of irradiations, Lacy and Leary (1958) studied the dimensional stability of alloys containing from 7 to 22 wt% (3 to 10 at.%) uranium. In this range, they found no variations in behaviour due to composition. For fuel-centre temperatures below approximately 550°C, the volume increases averaged 4% per atomic percent burnup of all atoms: At the highest burnup investigated, 3 at.%, there was a slight suggestion of an increase in the rate. By 800°C, substantial swelling (20% volume increase per atomic percent burnup) was observed. Willis (1959) found the swelling in $(\gamma_2 + \varepsilon)$ or in γ_2 greater than in $(\delta + \varepsilon)$, and greater still on cycling through the transformation.

Zirconium-uranium alloys, containing 3, 7, or 14 wt% uranium, that had been exposed at about 50°C to around 0.5, 1.0, and 2.5 at.% burnup, respectively, were subsequently subjected to successive isochronal anneals up to 900°C (Loomis, 1964). The resulting density changes per atomic percent burnup are plotted in Fig. 5–11. The release of

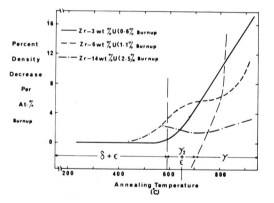

Fig. 5–11. The density decrease per unit burnup of zirconium-uranium alloys during post-irradiation annealing as a function of the annealing temperature. Derived from the results of Loomis (1964).

fission-product gas during annealing was also measured: Although all specimens released approximately the same *fractional* amount at 900°C, appreciable release started at 550°C from the Zr-14wt%U (high burnup), at 600°C from the Zr–6wt%U (intermediate burnup), but not until

700°C from the Zr–3wt%U (low burnup). Unfortunately, the composition and the atomic percent burnup vary together in this series of irradiations. From the shapes of the swelling curves, Loomis suggested that phase changes were having an important effect on the swelling character-istics. However, the appreciable swelling starting about 450°C in the two higher alloys indicates that the trans-formations alone do not fully explain the swelling. It was notable that, above 700°C, the highest alloy had the lowest density decrease per atomic percent burnup, and vice versa.

Earlier, Johnston (1958) had annealed alloys containing 7 and 8 wt% uranium that had been irradiated at 330°C up to 1.25% burnup of all atoms. Over 1% burnup, he observed nearly 10% decrease in density at an annealing temperature of 510°C, and about 25% at 540°C. Studying the effects of transients, Johnston found that even a few seconds annealing at temperatures over 650°C were sufficient to promote gross swelling.

Metallography of Johnston's specimens with 1% burnup after various anneals showed how porosity devel-oped with increasing temperature. At 425°C, the only pores visible were under 0.1-µm diam and were distributed through the zirconium matrix: At 510°C, some pores about 1-µm diam were apparent, associated with delta-phase particles of about the same size at grain boundaries, while the pores in the matrix had probably enlarged somewhat: And at 595°C, there was gross and irregularly shaped porosity, several microns across, outlining the matrix grains in which some of the finer pores still remained.

In these specimens, the uranium-rich particles were so fine (1- to 2-µm diam) that most of the damage due to recoils occurred in the matrix. Since Buckley (1962) has demonstrated that fission-fragment damage causes ir-radiation growth in anisotropic alpha zirconium (just as in alpha uranium), larger delta-phase particles would seem desirable. The metallographic structure of unirradiated alloys in this composition range can be profoundly altered by adjusting the heat treatment, but none of the products illustrated by Bauer (1959) in a review of these materials bore much resemblance to an ideal dispersion fuel (Sec. 5.1). Thus, the difference in swelling results obtained by Loomis and Johnston may be due to unknown variations in the initial structure of their specimens.

There are reasons for suspecting that other favourable results on the irradiation of zirconium-uranium alloys have not been published. For low neutron absorption, good corrosion resistance, and ease of bonding to zirconium sheathing, these alloys have an obvious potential in water-cooled power reactors. Indeed, an alloy based on Zircaloy and containing 6.33wt% uranium was used in the seed elements for the first core of the Shippingport Reactor, where a burnup of about half the uranium atoms was probably achieved, i.e., very approximately 1.5% of all atoms. If the results of Lacy and Leary are applicable, an increase in plate thickness of about 6% would have been expected in Shippingport. The absence of reports on the development or performance of the Shippingport seed elements suggests that information on these alloys is being restricted.

The physical and mechanical properties of zirconium-uranium alloys are affected by irradiation in qualitatively the same way as are those of other dispersion fuels such as aluminum-uranium alloys. For instance, the energy absorbed in fracturing bend-test specimens had decreased tenfold following irradiation to a burnup of 1% of all atoms at 350 to 650°C (Mehan, 1958).

Specimens of Zr–50wt%U, delta-phase UZr_2, irradiated at temperatures under 600°C, showed a linear increase in volume with increasing exposure up to a value of less than 15% at 4% burnup of all atoms (Eckel et al., 1956). Under 500°C, zirconium-uranium containing 40, 50, or 60 wt% uranium (the first and last compositions have some free zirconium and uranium, respectively, in the UZr_2) experienced a similarly low swelling rate, but the maximum burnup was only 0.4% of all atoms (Fillnow, 1956). A specimen of Zr–50wt%U was irradiated for 75 days in pressurized water with a deliberate hole in its sheath:

Although the fuel-surface temperature was 365°C, corrosion of the fuel was negligible and dimensional changes were no greater than for the intact elements (Fillnow).

If the fissionable material has to be dispersed for a particular reactor design, use of a moderator as a dispersant avoids the introduction of purely parasitic material. Thus, the carbon and BeO on coated particles (Sec. 5.5) serve a dual function and do not spoil the neutron economy. Similarly, zirconium hydride is a possible moderator in which uranium can be incorporated to yield a dispersion fuel. One advantage of such a system is that any temperature excursion in the fuel also affects the moderator and, by increasing the neutron temperature, decreases the reactivity: This "prompt negative temperature coefficient of reactivity" constitutes an inherent safety feature that makes zirconium-uranium hydride an attractive fuel for unattended reactors and for those used in training programmes. The possibility of hydrogen evolution due to decomposition of the fuel during a temperature excursion is less attractive.

In the zirconium-hydrogen system, a face-centered cubic phase is stable for compositions around $ZrH_{1.7}$. For H/Zr atomic ratios below about 1.4, the alpha-zirconium phase, saturated with hydrogen in solution, coexists with the hydride phase. When an alloy of Zr–8wt%U is hydrided to an H/Zr atomic ratio of 1.7, the uranium is precipitated in the zirconium hydride either as uranium metal or as a uranium-zirconium alloy. Specimens of (ZrH)–8wt%U and $(ZrH_{1.5})$–8wt%U were irradiated to 4% burnup of the uranium at surface temperatures of 10°C without any visible change in dimensions resulting (Wallace et al., 1958). In other tests, $(ZrH_{1.7})$–7wt%U was irradiated to approximately 1% burnup of the uranium at fuel temperatures between 630 and 740°C without obvious damage (Vetrano, 1960). Lamale et al. (1959) achieved much higher values of the burnup, 13 to 15% of the uranium, with fuel-centre temperatures from 480 to 760°C but in compositions containing much less uranium, $(ZrH_{1.65})$–2wt%U. The fuel pellets suffered marginally significant decreases in length and increases in diameter with the net result of a density decrease about 1%. Post-irradiation metallography showed the fuel to be single-phase material containing little porosity. However, no truly comparable unirradiated controls had been retained, so it is not known to what extent the initial material had contained uranium-bearing precipitates and porosity. Simnad et al. (1964) have reported that $(ZrH_{1.7})$–8.5wt%U irradiated to 12% burnup of the uranium at temperatures up to 650°C showed no significant change in density or metallographic structure.

Although the limitations of this fuel have clearly not yet been established, certain problems in its application can be anticipated. At any particular temperature a given composition is in equilibrium with a specific partial pressure of hydrogen. The pressure is probably insufficient to have any serious mechanical effect on the sheath but, unless the sheath is highly impermeable to hydrogen, the partial pressure of hydrogen in the coolant could eventually control the fuel composition, Also, the hydrogen is highly mobile at normal operating temperatures so that thermal diffusion of the hydrogen within the fuel could cause distortions. It is known that further work has been done on zirconium-uranium hydride fuels in connection with the SNAP type of reactors in the US space programme, and that the results are not published in the open literature.

5.9 Niobium-Uranium

At temperatures around 1000°C, niobium and gamma uranium, which both have the same face-centered cubic structure, form a continuous solid solution: Even at room temperature, niobium can accept over 20 at. % (40 wt%) uranium in solid solution. Niobium-rich niobium-uranium alloys are therefore of special interest in having the fissile atoms dispersed individually in a strong, isotropic metal of high symmetry. Unfortunately, the irradiation behaviour of these alloys has not been extensively investigated.

Specimens containing 10 and 20 wt% (4 and 9 at.%) uranium, irradiated up to 1% burnup of all atoms, showed volume increases of 2 or 3% per 1% burnup for calculated centre temperatures of 650°C, and about 4% per 1% burnup at 1000°C (DeMastry et al., 1963). There is no report of metallographic examination to investigate the nature of the porosity in these specimens, which remained smooth, straight, and undistorted during their irradiation. Other specimens containing 30 wt% (14 at.%) uranium, or 20 wt% uranium plus 10 or 20 wt% zirconium, showed appreciably greater swelling; about 4% volume increase per 1% burnup at 650°C, and 10 to 30% at 1000°C. The zirconium-bearing specimens had surface cracks, but these were probably caused by oxygen contamination from the surrounding sodium embrittling a layer subject to considerable extension.

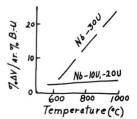

Post-irradiation mechanical testing showed that some hardening had occurred in the binaries exposed at the lower temperatures, but not at the higher ones. The ductility in bending was similarly affected, being zero for the lower temperatures and almost unchanged for the higher ones. Point defects produced by irradiation are presumably annealing rapidly by 1000°C. In the ternaries, the ductility was zero for both temperatures.

Isotropic niobium-uranium alloys would not be expected to exhibit the deformations and accelerated swelling around 450°C observed in uranium and attributed to irradiation growth. At higher temperatures, the exceptionally low swelling of the niobium-rich alloys may be a consequence of the absence of phase changes on thermally cycling between shutdown and operating temperatures, although low diffusion rates may also be contributing. The poorer resistance to swelling of the Nb–30wt%U alloy cannot be associated with structural changes, since this composition is still within the solid solubility range. Provided there was no incidental variation in the amount and distribution of precipitated impurity phases in the different compositions (Sec. 2.7), the answer may lie in increasing diffusion rates with increasing alloying addition. Further investigation of this system might yield a better understanding of the whole swelling phenomenon.

5.10 Alloys Containing Plutonium

With the possible exception of aluminum-plutonium alloys, which will be considered separately, the irradiation experience with alloys containing plutonium is too fragmentary to permit the drawing of firm conclusions. At best, some comparisons with the performance of the uranium-containing analogue may be obtained. However, the difficulty of obtaining reproducible results with uranium alloys warns against placing too much weight on a single, unconfirmed observation.

Since an attractive combination for fast reactors would be plutonium as fissionable material with natural uranium as fertile material, several irradiations of uranium-plutonium alloys have been performed. Uranium-molybdenum-plutonium ternaries have also been tested, following the good irradiation behaviour of uranium-molybdenum binaries (Sec. 2.7). Between 10 and 20% plutonium is soluble in the alpha and beta phases of uranium: At higher plutonium contents, the intermetallic compound UPu can be present at temperatures below 600°C.

Specimens containing from 3.7 to 18.7 wt% plutonium in uranium, irradiated with centre temperatures up to 490°C, demonstrated that both irradiation growth and surface roughening can occur in these alloys, just as in unalloyed uranium (Kittel and Paine, 1958). Swelling was found in four specimens, containing 7.5 or 10 wt% plutonium, that were irradiated unrestrained in sodium at temperatures from 315 to 700°C (Waldron et al., 1958). Although the burnup was low, approximately 0.3% of all atoms, volume increases of 10 to 25% were observed. The specimens with 10% plutonium exhibited slightly worse behaviour than the others, even at the lower temperatures. However, a suspicion of plutonium segregation in these latter alloys means that the results may be unrepresentative of ideal material.

Partly because of similar suspicions with regard to uranium-molybdenum-plutonium alloys irradiated at the same time as the binary alloys, a second series of the ternaries was tested (Frost et al., 1962b). Two molybdenum contents, 7 and 12 wt% (15 and 25 at.%), were combined in turn with two plutonium contents, 7.5 and 15 wt%, to provide four compositions each of which was exposed to two levels of burnup, about 8000 and 15 000 MWd/tonne (0.85 and 1.6% burnup of the fissionable atoms). All were carefully homogenized. The fuel-surface temperature during irradiation was 600°C, and the power output was about 150 W/cm length. With one exception, the swelling was roughly proportional to the burnup and lay in the range of 25 to 70% volume increase per 1% burnup of all atoms. Only a U–12wt%Mo–15wt%Pu specimen showed a lower swelling of 11.6% after 8720 MWd/tonne (0.8% burnup of all atoms). Optical metallography revealed that the swelling was due to the development of porosity throughout the specimens. The pores at grain boundaries were the first to enlarge and a grain-boundary network, probably of oxides and carbides, had been seen in un-irradiated samples. Although these results are discouraging when compared with those for uranium-molybdenum alloys, there remains the possibility that better control of sample purity might reduce the swelling. However, some of these specimens were close in composition to others irradiated by Mustelier (1962), about 17.5 wt% plutonium with 9, 11, or 12 wt% (18, 22, or 26 at.%) molybdenum in uranium, that exhibited a similar behaviour. His were tested with surface temperatures from 410 to 525°C, and exposed to 4400 to 8300 MWd/tonne: Volume increases from 30 to 105% were observed.

Molybdenum is one of the relatively abundant fission products that, along with zirconium, technetium, ruthenium, rhodium, and palladium, are only partially removed by pyrometallurgical methods for decontaminating spent fuel. Since these elements would therefore accumulate in highly irradiated fuel, synthetic alloys prepared from their naturally occurring isotopes have been tested. The term "fissium" (Fs) is conveniently used to describe the conglomeration of fission products, but it should be appreciated that there is no unique, generally agreed composition.

Indeed, in some contexts, fissium is intended to include volatile fission products that would be removed by pyrometallurgy, while in others it excludes zirconium which can be separated in an additional stage.

Specimens of uranium-fissium-plutonium, all with 10wt% fissium, were irradiated unrestrained at 400°C to 1.1% burnup of all atoms (Foote, 1962). Only the one containing 10wt% plutonium showed tolerable performance with a volume increase of 5.5%: Those containing 20 or 25wt% plutonium were distorted and had increased in volume by 50% or over. However, specimens of U–10wt%Fs–20wt% Pu survived a burnup of 1.6% of all atoms at approximately 650°C when restrained by strong sheaths, although the fuel had swollen about 15% to occupy initial clearances within the sheath. Horak et al. (1962a) tested the same alloy, but unrestrained, to exposures ranging from 1 to 1.8% burnup of all atoms. They found that below 370°C, swelling was small, less than 10% volume increase, but that it rapidly grew worse as the temperature was increased to 440°C. Kittel and Beck (1963), who irradiated the same alloy at 670°C, showed that a refractory metal sheath with a thickness-to-diameter ratio of 1:16 could satisfactorily confine swelling within the sodium annulus provided, up to a burnup of 2.5% of all atoms, although a defective sheath permitted localized swelling at a burnup as low as 0.3%.

These irradiations have been most useful in indicating some of the factors that may be important. Obviously it would be valuable to establish the minimum pressure necessary to restrain swelling to any arbitrary level: An upper limit to the volume increase is probably that volume in which all the gaseous fission products can be confined at the temperature of the fuel without exceeding the coolant pressure, but an appreciably lower value would be expected if the gas could be kept in small pores (Sec. 2.5). In the binary alloys, uranium-molybdenum, irradiation can affect the equilibrium phase structure (Sec. 2.6), and, presumably, a similar phenomenon occurs in both the uranium-molybdenum-plutonium ternaries and in the

fissium-containing analogues. If so, the irradiation behaviour could be profoundly affected by the fissioning
rate, but the possibility has not been systematically investigated. Even the equilibrium structures in the absence
of irradiation are poorly known for most compositions
of these ternaries.

In the thorium-plutonium system there is considerable
solubility of plutonium in the face-centred-cubic thorium.
Fuels containing up to approximately 20 wt% plutonium
are single phase, unlike their thorium-uranium counter-
parts in which uranium particles are dispersed in thorium.
The irradiation behaviour of these plutonium-bearing
alloys might therefore be expected to be very similar to
that of niobium-rich niobium-uranium alloys. However, a
specimen containing 15 wt% plutonium irradiated at 540°C
to a burnup of 0.54% of all atoms suffered a volume
increase of 14% (Waldron et al., 1958). Despite this
relatively large change, the specimen was undistorted and
its surface had a good appearance. Other specimens
containing 5 and 10% plutonium were irradiated at 450°C
to a burnup of 1.9 and 2.6% of all atoms, respectively
(Horak et al., 1962b). These, too, exhibited good appearance and their volume changes were only about 1% per 1%
burnup. Although the latter results are in line with expectations, the much larger swelling reported by Waldron
et al. has not been explained. The difference in swelling
is larger than would be expected for a temperature difference of only 100°C in a single-phase system without
transformations.

The solubility of plutonium in alpha zirconium is about
23 wt% (10 at.%). At higher temperatures, there is continuous solid solubility between beta zirconium and
epsilon plutonium, both body-centred-cubic, but for
plutonium contents over 23 wt%, at least two transformations could be involved in cooling the alloy to room
temperature. Specimens containing 5 and 7 wt% plutonium were irradiated at 500°C to a burnup of 1.4 and
1.9% of all atoms, respectively (Horak et al., 1962b).
The specimens had elongated to about three times their

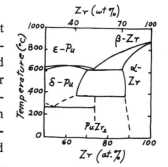

original length. They had been cold worked and were presumably highly textured, but it was not determined whether the growth was due to irradiation, *per se*, or to associated thermal cycling: Alpha zirconium is strongly anisotropic and therefore susceptible to both forms of growth (Secs. 2.2 and 2.3). Volume changes in these specimens were from 1 to 3.3% per 1 at.% burnup. An extruded alloy of higher plutonium content, 63 wt% (40 at.%), decreased in volume by 5.4% when irradiated at 500°C to a burnup of 0.83% of all atoms but suffered no serious elongation or surface roughening (Waldron et al., 1958). At its operating temperature, this specimen should have had a single-phase, body-centred-cubic structure, which would probably have been retained during the four periods that the temperature fell to 100°C at reactor shutdown.

5.11 Aluminum-Plutonium

The behaviour of aluminum-plutonium alloys is generally similar to that of the corresponding aluminum-uranium alloys, but no irradiations designed to give direct comparisons have been performed. For the compositions normally used, i.e., below 25 wt% (3.6 at.%) plutonium, the structure consists mainly of $PuAl_4$ precipitated in a matrix of nearly pure aluminum, although some $PuAl_3$ may also be observed. $PuAl_4$ has a similar defect structure to that of UAl_4 (Sec. 5.6).

Compositions up to at least 20 wt% plutonium can tolerate high exposures without significant swelling or distortion when irradiated at low temperatures. Loomis (1963) exposed Al–5wt%Pu to 89.3% burnup of the plutonium and Al–10wt%Pu to 75.6% burnup at temperatures under 100°C without decreasing the density by as much as 1%. Full-sized fuel elements containing up to 20 wt% plutonium were irradiated to a burnup of 55% of the plutonium with fuel-surface temperatures probably below 100°C (Runnalls, 1958): Although no post-irradiation measurements were made, a density decrease as high as 5%, had it occurred, would have been apparent.

Other experiments have shown good dimensional stability at higher temperatures. Jones (1962b) irradiated specimens containing approximately 5, 10, 15, and 20 wt% plutonium to exposures ranging from 7 to 20% burnup of the plutonium at fuel temperatures between 300 and 400°C. The measured increase in fuel volume rose linearly with exposure to a value of 2% at a burnup of 0.25% of *all atoms*, independent of composition when the burnup is expressed in these units. Such a swelling rate is large compared with that expected from solid fission products alone, and this suggests that gas-filled porosity may be contributing to the volume changes. These specimens were heavily restrained: They were sheathed in thick Zircaloy and cooled by pressurized water at 120 atm. Companion specimens were irradiated with deliberate clearances between the fuel and the sheath so that they were, at least initially, unrestrained and operating with fuel temperatures from 500 to 600°C. The swelling was increased by a factor of about 3.

Some other specimens containing 2 to 10 wt% plutonium (Jones, 1962a) exposed with the fuel at temperatures from 430 to 500°C and only lightly restrained, showed less than 0.3% increase in volume, but the burnup was only 6% of the plutonium (up to 0.5×10^{-3} of all atoms). Bailey et al. (1962) irradiated restrained specimens similar to those of Jones in coolant at 300°C and observed no serious expansions after a burnup of 20% of a 2.53 wt% plutonium content.

Jones (1962b) measured large increases of hardness in his cast specimens as a result of the irradiation, and both he and Loomis followed the recovery in post-irradiation annealing. Loomis's specimens, which were hot-rolled plates still in their aluminum sheathing, exhibited a single, sharp annealing stage around 300°C with no noticeable change at temperatures below or above up to 600°C. The specimens irradiated by Jones at temperatures over 300°C showed only a small decrease in hardness up to

580°C, then a large drop to values near or below the as-cast ones at 630°C. One of his specimens that was unrestrained, and at 500 to 625°C throughout, softened considerably during irradiation.

From studies of control specimens, it appeared that the 300°C stage represented primarily recovery of cold work in the aluminum: The slow decrease from 430 to 580°C was probably due to spheroidization of the fissile particles. After the 630°C anneal, some of Jones's specimens showed appreciable swelling, but most did not suffer large increases in volume, i.e., 15 to 35%, until 650°C. Thus, the drastic softening observed at high temperatures does not appear to be due to the porosity developed or even to be associated directly with the swelling. In both sets of alloys, the intermetallics were on a sufficiently fine scale that the fission products should have been fairly uniformly dispersed throughout the fuel — contrary to one of the principles of dispersion fuels. Thus, the softening may have been due to annealing of recoil damage in the matrix.

A complete loading of aluminum-sheathed Al–14wt%Pu fuel plates operated successfully in the Materials Test Reactor, where the coolant is cold water. In the Plutonium Recycle Test Reactor, with coolant temperatures up to 310°C, approximately 1500 Zircaloy-sheathed Al–2wt%Ni–1.8wt%Pu fuel rods performed well: The maximum burnup achieved was 83% of the plutonium.

Brief mention of plutonium-rich plutonium-aluminum alloys is included in Sec. 2.7.

6

Concluding Remarks

Although each fuel behaves differently under irradiation, qualitative similarities can be discerned. Indeed, a few general effects account for most of the irradiation behaviour observed.

When measured after an irradiation at relatively low temperature, i.e., well below $\frac{1}{2}\Theta_m$, where Θ_m is the material's melting point on the absolute temperature scale, many of the properties are appreciably affected by exposures as low as 10^{16} fissions/cm^3. For instance, certain changes in the electrical resistivity of uranium had saturated by 4×10^{16} fissions/cm^3, while the thermal conductivity, lattice parameter, and hardness of UO_2 and the electrical resistivity and lattice parameter of UC all showed significant changes by 10^{16} fissions/cm^3.

These properties are sensitive to the presence of point defects, and individual vacancies and interstitials probably cause much of the damage at low exposures. However, electron microscopy has given direct evidence for more complex forms of damage in several fuels. Interstitials seem to be the first to form clusters, which then grow into dislocation loops and finally dislocation tangles as the exposure is increased. These clusters and dislocations, by impeding the motion of other dislocations, are presumably responsible for increasing the yield strength and decreasing the ductility of uranium, thorium, and alloys such as aluminum-uranium. Subsequent annealing, or irradiation at higher temperatures, modifies the irradiation damage without necessarily removing it altogether, as shown by electron microscopy and conductivity measurements on UO_2 and UC.

As the exposure is increased, the increase in lattice parameter of UO_2 passes through a well-defined and reproducible maximum about 10^{17} to 10^{18} fissions/cm^3. The effect can probably be attributed to a depletion of single interstitials as the interstitial loops grow. However, it is notable that in UC, which is the only other material for which changes in lattice parameter have been studied to a comparable extent, no such maximum has been observed. The more strongly ionic bonding in UO_2 may be affecting the clustering of individual point defects, but many further studies will be required to identify the particular defects responsible for specific effects. The work by Amelinckx on uranium oxide and by Childs on uranium carbide indicates what can be achieved, and also illustrates the need for close control of the experimental conditions. Preferably, the same batch of specimens, all irradiated at the same time, would be used for the measurement of several different properties and for examination by electron microscopy.

Electron microscopy reveals pores in fissile materials for exposures in excess of roughly 10^{18} fissions/cm^3, but their existence at much lower exposures, at least after a post-irradiation anneal, can be inferred from studies of gas release from UO_2 and possibly UC. At temperatures about $\frac{1}{2}\Theta_m$, and over, the small pores are appreciably mobile. The migration of pores, which appears to be a universal phenomenon for irradiated fuels, is of great significance to both gas release and swelling. Dislocations, precipitate particles, and grain boundaries have been shown to interact strongly with the pores. Thus, pores are trapped on grain boundaries and, by colliding with others there, grow more rapidly than they would otherwise.

A thermal gradient has been demonstrated to be one of the factors providing an effective driving force for moving the pores. The major release of fission-product gases from solid UO_2, while uranium largely retains the gases to its melting point, may be associated with the higher thermal gradients normal in UO_2. However, UC, with a thermal conductivity closer to that of uranium than UO_2, has gas-release characteristics very similar to those of UO_2: Indeed, it will take more carefully controlled experiments than have yet been performed to prove a difference between the gas release from UC and UO_2 under the same conditions.

The tendency for swelling is in the opposite direction to that for gas release. Swelling generally becomes apparent at a burnup between 0.1 and 1% in uranium, and at several percent burnup of the uranium in UC and UO_2. Again the similarity between the last two, despite the difference in their conductivities, is a reminder of how much has still to be understood concerning pore behaviour under irradiation and at elevated temperatures.

In fuels that are anisotropic, high internal stresses, due to thermal expansion around fission spikes, direct vacancies and interstitials to cluster on different crystallographic planes. This form of segregation causes irradiation growth and, consequently, certain types of swelling in uranium. The growth observed in irradiated zirconium-plutonium alloys, which are anisotropic, is probably due to the same cause. If an external stress is applied during irradiation, the internal stresses can greatly accelerate the normal creep rates so that anisotropic fuel generally deforms readily under irradiation. Such an effect has been observed in ZrO_2–UO_2, as well as in uranium. Even in isotropic UO_2, irradiation can enhance creep rates, but less than in ZrO_2–UO_2. The creep behaviour under irradiation of isotropic fuels now merits further attention. In comparing the irradiation behaviour of various fuels, the nature of the atomic bonding (ionic, covalent, or metallic) appears to be less important than whether or not the crystal structure is isotropic.

Irradiation destroys the crystalline structure of some anisotropic fuels, e.g., U_3O_8 and Al_2O_3–UO_2, but annealing at a suitably elevated temperature restores the structure. Irradiation can also promote atomic disordering that would not occur at the same temperature without irradiation. Initial variations in composition through the fuel tend to be levelled out by the fission spikes during irradiation: Studies of uranium-molybdenum and uranium-niobium alloys and of ZrO_2–UO_2 have shown that, where the original structure of a two-phase material is sufficiently fine, the homogenization can be important.

In dispersion fuels the effects in the fissile phase are largely those seen in the equivalent bulk material, while those in the matrix are due primarily to fast neutron damage. At the interface, matrix damage by recoiling fission fragments (steel–UO_2 cermets and carbon-coated particles) and irradiation-enhanced chemical reaction (aluminum–UO_2 cermets) have been observed. Where the fissile material is distributed in an isotropic matrix on a scale small compared with the recoil range, e.g., thorium-uranium and niobium-uranium alloys, the swelling rates are similar and relatively low for irradiation temperatures up to $\frac{1}{2}\Theta_m$.

The qualitative similarities between various fuels should not be allowed to obscure the many quantitative differences that are extremely important in practice. Also, it must be stressed that, even with a given fuel, minor differences in impurity content, pre-irradiation treatment, and conditions during irradiation can have a profound effect. The fact that the properties of irradiated fuel depend on its irradiation history, and not just on the final temperature and exposure, makes most complicated adequate description of the irradiation conditions.

7

References

	Page Numbers
Adam J. and Cox B., "The Irradiation-Induced Phase Transformation in Zirconia Solid Solutions," *Reactor Sci. Technol*, **11**, 31 (1959).	196
Adam J. and Rogers M.D., "X-Ray Diffraction Studies of Fission Fragment Damage in Uranium Carbide and Nitride," *Reactor Sci. Technol.*, **14**, 51 (1961).	210, 211, 226
Adam J. and Rogers M.D., Unpublished work quoted by Frost, (1963).	213
Adda, Y., Mustelier J.-P. and Quéré Y., "Behaviour of Uranium under Irradiation," *3rd. ICPUAE*, P/62, United Nations, (1964).	44, 69, 82, 91
Amelinckx S. and colleagues at Mol, "Physical Properties of UO_2 Single Crystals," Euratom Report EUR–2042, (1965).	117, 174
Anderson R.G. and Bishop J.F.W., "The Effect of Neutron Irradiation and Thermal Cycling on Permanent Deformations in Uranium under Load," in *Uranium and Graphite*, p. 17, Inst. of Metals, London, (1962).	37
Armstrong W.M., Irvine W.R. and Martinson R.H., "Creep Deformation of Stoichiometric Uranium Dioxide," *J. Nucl. Mater.*, **7, 2**, 133 (1962).	133
Auskern A., "Release of Xenon from Uranium Carbide at Low Temperatures," *J. Amer. Ceram. Soc.*, **47**, 8, 390 (1964).	140, 220
Auskern A. and Osawa Y., "Xenon Diffusion in Uranium Carbide Powder," *J. Nucl. Mater.*, **6**, 3, 334 (1962).	219, 220
Bailey W.J. and Chikalla T.D., "Irradiation of Uranium-Plutonium Oxide," US Report HW–SA–3129, (1964). Abstract in *Trans. Amer. Nucl. Soc.*, **6**, 2, 350 (1963).	185, 186
Bailey W.J., Freshley M.D. and Sharp R.E., "The Irradiation Testing of a Plutonium Fuel Element Concept at Hanford," *Trans. Amer. Nucl. Soc.*, **5**, 1, 252 (1962).	279
Bain A.S., "The Heat Rating Required to Produce Central Melting in Various UO_2 Fuels," ASTM Spec. Tech. Publicn. 306, 30 (1961)	176
Bain A.S., "Cracking and Bulk Movement in Irradiated Uranium Oxide Fuel Elements," Canadian Report AECL–1827, (1963).	106, 107, 109, 132, 163
Bain A.S., Private communication, (1964).	195
Bain A.S., "Highly Rated Hollow UO_2 Pellets Irradiated in Collapsible Sheathing—Dimensional Stability, Metallography and Autoradiography," Canadian Report AECL–2276, (1965).	110
Bain A.S., Christie J. and Daniel A.R., "Trefoil Bundles of NPD 7–Element Size Fuel Irradiated to 9100 MWd/tonne U," Canadian Report AECL–1895, (1964).	113

Bain A.S., MacDonald R.D. and Murray A.D., Private communication, (1964). 165

Bain A.S., May J.E. and Morel P.A., "UO$_2$/Sheath Heat Transfer in Large Diametrical-Clearance Elements Irradiated for Short Duration," *Trans. Amer. Nucl. Soc.*, **7**, 2, 433 (1964). 137

Bain A.S., Robertson J.A.L. and Ridal A., "UO$_2$ Irradiations of Short Duration—Part II," Canadian Report AECL–1192, (1961). 134

Bainbridge J.E. and Hudson B., "On the Techniques for Observing Fission Gas Bubbles in Uranium," submitted to the *J. Nucl. Mater.*, (1965). 55

Ball J.G., "Metallurgical Research in Nuclear Power Production," *J. Inst. Metals*, **84**, 239 (1956). 80

Ballif J.L., "Cladding and Bonding Materials Development for Advanced Organic Moderated Reactor—Progress Report," US Report NAA–SR–7777, III–D–11, (1962). 89

Barnes R.S., "Fundamentals of Inert-Gas Agglomeration and Swelling in Metals," in *Nuclear metallurgy—Vol. VI*," 21, AIME Publicn., (1959). 53

Barnes R.S., "A Theory of Swelling and Gas Release for Reactor Materials," *J. Nucl. Mater.*, **11**, 2, 135 (1964). 50, 60, 64, 73

Barnes R.S., Bellamy R.G., Butcher B.R. and Mardon P.G., "The Irradiation Behaviour of Uranium and Uranium Alloy Fuels," *3rd ICPUAE*, P/145, United Nations, (1964). 44, 45, 84, 86, 89

Barnes R.S. and Mazey D.J., "The Migration and Coalescence of Inert Gas Bubbles in Metals," *Proc. Roy. Soc.*, **275A**, 47 (1963). 53, 58, 59, 62

Barnes R.S. and Mazey D.J., "The Movement of Helium Bubbles in Uranium Dioxide," *3rd Proc. European Regional Conf. Electron Microscopy, Prague*," p. 197. 1964, 147

Barney W.K. and Wemple B.D., "Metallography of Irradiated UO$_2$-Containing Fuel Elements," US Report KAPL–1836, (1958). 151, 241, 243

Bates J.C., "A comparison of the Thermal Conductivity of Irradiated and Unirradiated Uranium," UK Report R&DB(W)TN–78, (1953, declassified 1958). 90

Bates J.L., "Microhardness of Uranium Dioxide," US Report HW–77799, (1963). 170, 171

Bates J.L., "Radiation Damage to UO$_2$," in "Ceramics Research and Development Operation Quarterly Report July-Sept. 1963," US Report HW–76303, p. 2.22, (1964). 168

Bates J.L., "Metallic Uranium in Irradiated UO$_2$," US Report HW–82263, (1964). 164

Bates J.L., "Fission Product Distribution in Irradiated UO$_2$," *Trans. Amer. Nucl. Soc.*, **7**, 2, 389 (1964). 160

Bates J.L., Christensen J.A. and Daniel J.L., "Irradiation Effects in Uranium Dioxide Single Crystals," US Report HW–73959, (1962). 168

Bates J.L., Christensen J.A. and Roake W.E., "Fission Products and Plutonium Migrate in Uranium-Dioxide Fuel," *Nucleonics*, **20**, 3, 88 (1962). 160, 186

Bauer A.A., "An Evaluation of the Properties and Behavior of Zirconium-Uranium Alloys," US Report BMI–1350, (1959). 270

Bauer A.A., Kass S. and Goldman K.M., "Physical Metallurgy and Properties of Zirconium-Uranium Alloys," *2nd ICPUAE*, **5**, 602, United Nations, (1958). 85

Bellamy R.G., "The Swelling of Alpha-Uranium under Neutron   41, 43, 85
Irradiation to 0.7% Burnup," in *Uranium and Graphite*, p. 53.
Inst. of Metals, London, (1962)

Bellamy R.G., Private communication, (1965): Several of the results   43, 44
are given by Barnes et al., (1964).

Belle J. "Uranium Dioxide: Properties and Nuclear Applications,"   140, 150
US Govt. Printing Office, Washington, DC, (1961).

Belle J. and Lustman B., "Properties of UO_2," US Report WAPD–   143
184, (1957).

Bement A.L., "The Effects of Low Neutron Exposures at Low Temper-   91
atures on the Hardness and Tensile Properties of Natural
Uranium," US Report HW–60326, (1959).

Bentle G.G., "Irradiation Swelling of Uranium and Uranium Alloys,"   41, 49, 83
US Report NAA–SR–4893, (1961).

Berman R.M., "Fission Fragment Distribution in Irradiated UO_2,"   165
Nucl. Sci. Eng., **16**, 3, 315 (1963).

Berman R.M., Private communication, (1965).   78

Berman R.M., Bleiberg M.L. and Yeniscavich W., "Fission Fragment   168, 189
Damage to Crystal Structures," *J. Nucl. Mater.*, **2**, 2, 129 (1960).

Biddle P. "The Emission of Xenon-133 from Uranium Carbonitrides,"   220, 228
UK Report AERE–R4700, (1964).

Bierlein T.K. and Mastel B., "Metallographic Studies of Uranium,"   93, 96
2nd ICPUAE, **6**, 14, United Nations, (1958).

Bierlein T.K. and Mastel B., "Damage in UO_2 Films and Particles   114
during Reactor Irradiation," *J. Appl. Phys.*, **31**, 2314 (1960).

Bierlein T.K. and Mastel B., "Fission-Induced Vaporization of UO_2 139
from a Source and Subsequent Condensation on Collectors Exposed
to Fission Fragment Bombardment," *J. Nucl. Mater.*, **7**, 1, 32 (1962).

Billington D.S., "Radiation Damage in Reactor Materials," *1st*   90, 91
ICPUAE, **7**, 421, United Nations, (1955).

Blank H. and Amelinckx S., "Defects in Irradiated UO_2 Single   114
Crystals," *J. Appl. Phys.*, **34**, 8, 2200 (1963).

M.L. Bleiberg, "A Kinetic Study of Irradiation Induced Phase Changes   77, 78
in Uranium–9 w/o Molybdenum Alloy," *Nucl. Sci. Eng.*, **5**, 2,
78 (1959).

Bleiberg M.L., Berman R.M. and Lustman B., "Effects of High   114, 115, 154, 168,
Burnup on Oxide Ceramic Fuels," in *Radiation damage in*   190, 191, 195, 198
reactor materials, p. 319, IAEA, Vienna, (1962).

Bleiberg M.L. and Jones L.J., "The Effects of Pile Irradiation on   229, 232
U_3Si," *Trans. AIME*, **212**, 758 (1958).

Bleiberg M.L., Jones L.F. and Lustman B., "Phase Changes in Pile   77, 79
Irradiated Uranium-Base Alloys," *J. Appl. Phys.*, **27**, 11, 1270 (1956).

Bleiberg M.L., Yeniscavich W. and Gray R.G., "Effects of Burnup   189, 190, 191
on Certain Ceramic Fuel Materials," ASTM Spec. Tech. Publcn.
306, p. 64. (1961)

Blin J., "Energy of Solution of Rare Gases in Metals," in Colloque   54
sur la Diffusion a l'État Solide organisé à Saclay les 3, 4 et 5 Juillet
1958, 65, CEN Saclay, (1958).

Bloch J., "Phase Change and Disordering of U_2Mo Caused by   76
Irradiation," *J. Nucl. Mater.*, **2**, 1, 90 (1960).

Bloch J., "Thermal Restoration of the Lattice Parameter of Irradiated 167
UO_2," *J. Nucl. Mater.*, **3**, 2, 237 (1961).

Bloch J., Mustelier J.-P., Bussy P. and Blin J., "Crystallographic   93
Studies of Irradiated Uranium," *2nd ICPUAE*, **5**, 593, United
Nations, (1958).

Blomeke J.O. and Todd, M.F., "Uranium–235 Fission-Product   8
Production as a Function of Thermal Neutron Flux, Irradiation
Time and Decay Time," US Report ORNL–2127, (1958).

Bogaievski M., Caillat R., Delmas R., Janvier J.C. and Robertson   124
J.A.L., "Direct Measurement of the Thermal Conductivity of
Uranium Oxide In-Reactor to 1200°C," in *New Nuclear Materials
Including Non-metallic Fuels*, Vol **I**, 307, IAEA, Vienna, (1963).

Boltax A., "A Metallographic Study of Swelling in Alpha-Particle   53
Irradiated Aluminum and Aluminum Alloys," *J. Nucl. Mater.*,
7, 1, 1 (1962).

Bomar E.S. and Gray R.J., "Thorium-Uranium Carbides for Coated-   253
Particle Graphite Fuels," in *Nuclear Metallurgy*, Vol. **X**, p. 703,
AIME Publicn., (1964).

Booth A.H. and Rymer G.T., "Determination of the Diffusion   141
Constant in UO_2 Crystals and Sintered Compacts," Canadian
Report AECL–692, (1958).

Boyko E.R., Eichenberg J.D., Roof R.B. and Halteman E.K., "X-Ray   173
Examination of Irradiated UO_2," US Report WAPD–BT–6,
(1958).

Bradbury B.T., Cole J.E., Frost B.R.T. and Lambert J.D.B., "Irradi-   203, 206, 225
ation Effects in U–C and (U, Pu)–C Fuel Pellets," in *Carbides in
Nuclear Energy*, Vol. **2**, p. 879, Macmillan and Co. Ltd., London,
(1963)

Bromley J., "Transport and Diffusion of Fission Products in Graphite," 8, 251
in *Progress in Nuclear Energy*, Series IV, **5,** 173 (1963).

Buckley S.N., "Irradiation Growth," in *Properties of Reactor   29, 33, 34, 35, 37
Materials and the Effects of irradiation Damage*, p. 413. Butterworths
London, (1961).

Buckley S.N., "The Relationship between Irradiation Growth in   29, 30, 31, 37, 269
Mono- and Polycrystalline α-U," in *Uranium and Graphite*,
p. 41. Inst. of Metals, London, (1962)

Buckley S.N., Unpublished work quoted by Pugh and Butcher, (1964).   27, 28

Buckley S.N., Private communication, (1965).   28

Buckley S.N., Harding A.G. and Waldron M.B., "Physical Damage   25
Brought about by Thermally Cycling Uranium through its Phase
Changes," *J. Inst. Metals*, **87**, 150, (1959).

Bugl J. and Keller D.L., "Uranium Mononitride—a New Reactor Fuel,"   227
Nucleonics, **22**, 9, 66 (1964).

Burger G., Isebeck K,. Wenzl H., Jousset J.C. and Quéré Y., "Irradi-   91
ation of Uranium with Neutrons at 4.5°K," *Phil. Mag.*, Series 8, **11**,
621 (1965).

Burian R.J., Miller N.E., Ritzman R.L. and Townley C.W., "Initial   256
Irradiation Evaluation of Experimental Pyrolytic-Carbon-Coated
UC_2 and UO_2 Fuel Particles," US Report BMI–1628, (1963).

Burkart, M.W. "The Corrosion Mechanism of Uranium Base Alloys in   98
High Temperature Water," Part III of "Development and Properties of Uranium-Base Alloys Corrosion Resistant in High Temperature Water," US Report WAPD–127, (1956).

Bush S.H., "Irradiation Effects in Uranium," US Report HW–51444,   91
(1957).

Carroll R.M., "Argon Activation Measures Irradiation Flux Continuously," *Nucleonics*, **20**, 2, 42 (1962). 20

Carroll R.M., "Fission-Gas Release From Single-Crystal UO_2 During Irradiation," *Trans. Amer. Nucl. Soc.*, **6**, 1, 121 (1963). 139

Carroll R.M. and Sisman O., "In-Pile Fission-Gas Release From Single-Crystal UO_2," *Nucl. Sci. Eng.*, **21**, 2, 147 (1965). 139

Childs, B.G.,"The Release of Stored Energy From Neutron-Irradiated Uranium Oxides," *J. Nucl. Mater.*, **5**, 1, 128 (1962). 172

Childs B.G. and Buxton K., "High-Precision Density Measurements on Irradiated Uranium Carbides," UK Report TRG–R862(D), (1965). 209, 210, 211, 212

Childs B.G., Ogilvie A., Ruckman J.C. and Whitton J.L., "The Low-Temperature Irradiation Behaviour of Cast Uranium Carbide," in *Radiation Damage in Reactor Materials*, p. 241. IAEA, Vienna, (1962). 204

Childs B.G. and Ruckman J.C., "Radiation Damage Effects in Uranium Carbide," in *New Nuclear Materials Including Nonmetallic Fuels*, Vol. II, p. 1. IAEA, Vienna, (1963) 204, 210, 211

Childs B.G., Ruckman J.C. and Buxton K., "Radiation Damage in UC, UC_2 and U_2C_3," in *Carbides in Nuclear Energy*, Vol. 2, p. 849. MacMillan and Co. Ltd., London, (1963) 204, 210, 211, 213

Christensen J.A., "Irradiation Effects on Uranium Dioxide Melting," US Report HW–69234, (1962). 119

Christensen J.A., "Thermal Expansion and Change in Volume on Melting for Uranium Dioxide, US Report HW–75148, (1963). 120

Christensen J.A., "Thermal Conductivity of Nearly Stoichiometric UO_2—Temperature and Composition Effects," US Report WCAP–2531, (1963). 121, 128

Christensen J.A. and Allio R.J., "In-Pile Measurement of Uranium Dioxide Fuel-Temperature Distribution," *Trans. Amer. Nucl. Soc.*, **8**, 1, 40 (1965). 129

Christensen J.A., Allio R.J. and Biancheria A., "Melting Point of Irradiated Uranium Dioxide," *Trans. Amer. Nucl. Soc.*, **7**, 2, 390 (1964). Further details in US Report WCAP–6065, 1965. 119

Churchman A.T. and Barnes R.S., "Swelling and Inert Gas Diffusion in Irradiated Uranium—Swelling of Irradiated Uranium on Subsequent Out-of-Pile Heating—Material Originally Irradiated to 0.3 to 0.4% Burnup," *2nd ICPUAE*, **5**, 548, United Nations, (1958). 72

Churchman A.T., Barnes R.S. and Cottrell A.H., "Effects of Heat and Pressure on the Swelling of Irradiated Uranium," *J. Nucl. Energy*, **7**, 88 (1958). 50, 67

Clayton J.C., "The Evolution of Xe–133 From Slightly-Irradiated Zirconia Urania Plates," US Report WAPD–TM–313, (1962). 197

Clough D.J., "Experiments to Determine the Effects of Burnup on the Effective Thermal Conductivity of Stoichiometric Sintered UO_2," UK Report AERE–R4146, (1962). 124

Clough D.J. and Sayers J.B., "The Measurement of the Thermal Conductivity of UO_2 under Irradiation in the Temperature Range 150°C–1600°C," UK Report AERE–R4690, (1964). 124, 126

Cohen I., Lustman B. and Eichenberg J.D., "Measurement of the Thermal Conductivity of Metal-Clad Uranium Oxide Rods during Irradiation," *J. Nucl. Mater.*, **3**, 3, 331 (1961). 125

Colabianchi G., Fizzotti C., Traversi G., Stobo J.J. and Squadrelli F., 125
"Irradiation Experiments on Uranium-Rich Alloys," *3rd ICPUA*,
P/735, United Nations, (1964).

Crane J., Gordon E. and Gates J.E., "Effects of Fabrication and 202, 204, 207, 215,
Composition on the Irradiation Stability of Uranium Carbide," 216, 221, 223
in *Nuclear Metallurgy*, Vol. X, 765, AIME Publicn., (1964).

Daniel J.L., Matolich J. and Deem H.W., "Thermal Conductivity of 122, 127
UO_2," US Report HW–69945, (1963).

Daniel R.C., Bleiberg M.L., Meieran H.B., and Yeniscavich W., 152, 155, 156
"The Effect of High Burnup on Zircaloy-Clad Bulk UO_2 Plate
Fuel Element Samples," US Report WAPD–263, (1962).

Daniel R.C. and Cohen I., "In-Pile Effective Thermal Conductivity 125, 194
of Oxide Fuel Elements to High Fission Depletions," US Report
WAPD–246, (1964).

Davies D. and Long G., "The Emission of Xenon–133 From Lightly 141, 142
Irradiated Uranium Dioxide Spheroids and Powders," UK Report
AERE–R4347, (1963).

Davies D., Long G. and Stanaway W.P., "The Emission of Volatile 161
Fission Products From Uranium Dioxide," UK Report AERE–
R4342, (1963).

Dayton R.W., Goeddel W.V. and Harms W.O., "Ceramic Coated 252, 257
Particle Nuclear Fuels," *3rd ICPUAE*, P/235, United Nations, (1964).

Deem H.W., Poberskin M., Lusk E.C., Lucks C.F. and Calkins G.D., 91
"Effect of Radiation on the Thermal Conductivity of Uranium-
1.6 w/o Zirconium," US Report BMI–986, (1955).

de Halas D.R. and Horn G.R., "Evolution of Uranium Dioxide 129, 130
Structure During Irradiation of Fuel Rods," *J. Nucl. Mater.*, **8**, 2,
207 (1963).

de Mastry J.A., Bauer A.A. and Dickerson R.F., "How Irradiation 273
at 1100–1800°F Affects Niobium-Uranium Alloys," *Nucleonics*,
21, 10, 86 (1963).

Eckel J.F., Bruch C.A., Levy A., Willis A.H. and Seymour W.E., 270
"Radiation Effects on Reactor Materials—A Review of KAPL
Work," in US Report TID–7515 (Pt 2) (Del), 170, (1956).

Eldred V.W., "Swelling and Inert Gas Diffusion in Irradiated Uranium 72, 93
—Swelling of Irradiated Uranium on Subsequent Out-of-Pile
Heating—Material Irradiated to 0.17% Burnup at 300°C," *2nd
ICPUAE*, **5**, 551, United Nations, (1958).

Eldred V. W., Discussion recorded in *Properties of Reactor* 71
Materials and Effects of Irradiation Damage, Butterworths, London,
p. 540. (1961).

Eldred V.W., Greenough G.B. and Leech, P. "Fuel Element Be- 32, 33, 94
haviour Under Irradiation," *2nd ICPUAE*, **5**, 510, United Nations,
(1958).

Elleman T.S., Price R.B. and Sunderman D.N., "Fission-Fragment- 256
Induced Stresses in Ceramic Materials," US Report BMI–1635,
(1963).

Englander M. et al., "The Effect of Neutron Irradiation on Certain 46, 84, 93, 95
Fuels Used in Gas-Graphite Type Reactors," *3rd ICPUAE*, P/97,
United Nations, (1964).

Feraday M.A. and Chalder G.H., "A Collapse Mode of Failure in 178
Powder-Filled Fuel Elements," Canadian Report AECL–1894,
(1964).

Feraday M.A., Chalder G.H. and Martineau Y., Unpublished work referred to by Feraday and Chalder, 1964, (1965). 178

Field J.H., Leitz F.J., McNelly M.J. and Nelson R.C., "Apparent Boiling of Uranium Oxide in the Center of a Fuel Pin During Transient Power Generation," *Nucl. Sci. Eng.*, **14**, 210 (1962). 110

Fillnow R.H., "Summary of Bettis Irradiation Data on Reactor Fuel, Structural and Control Rod Materials," in US Report TID–7515 (Pt 2) (Del), 268, (1956). 270

Flaherty W.J., "Low Temperature Irradiation Sintering of Swaged UO_2—Progress Report," in US Report HW–76300, 4.11, (1963). 177

Flaherty W.J., Horn G.R. and Pember L.A., "Long-Term Irradiation of Cermet Fuels—Progress Report," in US Report HW–76304, 4.26, (1964). 250

Flaherty W.J. and Panisko F.E., "Short Term Irradiation of W–UO_2 Cermets—Progress Report," in US Report HW–76303, 4.35, (1964). 250

Foote F.G., "Plutonium Fuel Programs at Argonne National Laboratory," in US Report HW–75007, 7.1, (1962). 276

Fox A.C., Jackson E.E., Junkison A.R. and Wait E., "X-Ray Studies of Radiation Damage in Fissile Materials, Part II Uranium Dioxide" UK Report AERE–R4267, (1963), quoted by Roberts et al., 1964. 166

Freas D.G. Austin A.E. and Rough F.A., "Mechanism of Irradiation Damage in Uranium Monocarbide," in ASTM Spec. Tech. Publicn. 306, 131, (1961). 204, 205, 207, 210, 213

Freshley M.D. and Carroll D.F., "The Irradiation Performance of MgO–PuO_2 Nuclear Fuel," US Report HW–SA–3127, (1963). Abstract in *Trans. Amer. Nucl. Soc.*, **6**, 2, 396 (1963). 197

Freshley M.D. and Last G.A., "Irradiation of U–Mg Matrix Fuel Material to High Exposures," US Report HW–43973 (Del), (1956). 264

Freshley M.D. and Mattys H.M., "Irradiation of (Thorium, Plutonium)O_2," *Trans. Amer. Nucl. Soc.*, **7**, 2, 402 (1964). 187

Frost B.R.T., "The Carbides of Uranium," *J. Nucl. Mater.*, **10**, 4, 265 (1963). 200, 201

Frost B.R.T., Bradbury B.T. and Griffiths L.B., "Irradiation Effects in Fissile Oxides and Carbides at Low and High Burnup Levels," in *Radiation Damage in Reactor Materials*, p. 219, IAEA, Vienna, (1962). 203, 212

Frost B.R.T. Cope L.H., Lambert J.D.B. Lloyd H., Long W., Manson J.E. and Mardon P.G., "Fabrication and Irradiation Studies of UO_2-Stainless Steel Cermets," *3rd ICPUAE*, P/153, United Nations, (1964). 241

Frost B.R.T., Mardon P.G., and Russell L.E., "Research on the Fabrication, Properties and Irradiation Behaviour of Plutonium Fuels for the UK Reactor Programme," in US Report HW–75007, 4.1, (1962). 184, 275

Garlick A. and Shaw D., "The Thermal Conductivity of Irradiated Uranium," *J. Nucl. Mater.*, **16**, 3, 333 (1965). 90

Gates J.E., Lamale G.E. and Dickerson R.F., "The Examination and Evaluation of Irradiated Thorium—11 w/o Uranium Specimens," US Report BMI–1334, (1959). 266

Gavin A.P. and Crothers C.C., "Irradiation of an Aluminum Alloy-Clad, Aluminum-Uranium Alloy-Fuelled Plate," US Report ANL–6180, (1960). 262

Gerhart J.M., Siltanen J.N. and Cochran J.S., "The Irradiation and   184
Examination of a Plutonium-Uranium Oxide Fast Reactor Fuel,"
in ASTM Spec. Tech. Publicn. 306, 154, (1961).

Gibson G.W. and Francis W.C., "Annual Progress Report on Fuel   261
Element Development for FY 1962," US Report IDO–16799,
(1962).

Gilbreath J.R. and Simpson O.C., "The Effect of Reactor Irradiation   192
on the Physical Properties of Beryllium Oxide," *2nd ICPUAE*, **5**,
367, United Nations, (1958).

Gittus J., *Uranium*, Butterworths, London, (1963).   25

Goeddel W.V., "Pyrolytic-Carbon-Coated Carbidé Fuel Particles   252
and Their Use in Graphite Matrix Fuel Compacts," in US Report
TID–7654, 142, (1962).

Golyanov V.M. and Pravdyuk N.F., "Behaviour of Nuclear Fuel   108, 114
Under Irradiation—Investigation of Thin Layers of Irradiated
Uranium Dioxide," *3rd ICPUAE*, P/338, United Nations, (1964).

Goslee D.E., "Improving Performance of Stainless-Steel-UO_2 Cermet   240
Fuels," *Nucleonics*, **21**, 7, 48 (1963).

Graber M.J., Gibson G.W., Walker V.A. and Francis W.C., "Results   245, 247
of ATR Sample Fuel Plate Irradiation Experiment," US Report
IDO–16958, (1964).

Granata S. and Saraceno F., "The Relationship Between Burnup,   47
Temperature and Swelling in Alpha Uranium," *J. Nucl. Mater.*,
9, 3, 367 (1963).

Gray R.G. and Mrazik F.P., "Examination of an In-Pile Failure of   159
a PWR Core 1 Type Fuel Rod," in US Report WAPD–BT–23,
53, (1961).

Greenberg S., "Corrosion of Irradiated Uranium Alloys," *Nucl.*   99
Sci. Eng., **6**, 2, 159 (1959).

Greenberg S. and Draley J.E., "Effects of Irradiation on Corrosion   99
Resistance of some High Uranium Alloys," *Nucl. Sci. Eng.*,
3, 1, 19 (1958).

Greenwood G.W., "Volume Increases in Fissile Materials on Neutron   48
Irradiation and the Effects of Thermal Fluctuations," *J. Inst.*
Metals, **88**, 31 (1959).

Greenwood G.W., "Observations on Uranium Irradiated in the   75
α- and Subsequently in the γ-Phase," in *Properties of Reactor*
Materials and the Effects of Irradiation Damage, p. 475. Butterworths,
London, (1961)

Greenwood G.W., "The Effects of Neutron Irradiation on γ-Uranium 52, 71
and some Fissile Alloys of Cubic Crystal Structure," *J. Nucl.*
Mater., **6**, 1, 26 (1962).

Greenwood G.W. and Boltax A., "The Role of Fission Gas Re-   56, 69, 72
solution During Post-Irradiation Heat Treatment," *J. Nucl. Mater.*,
5, 2, 234 (1962).

Greenwood G.W. and Speight M.V., "An Analysis of the Diffusion   55, 60
of Fission Gas Bubbles and its Effect on the Behaviour of Reactor
Fuels," *J. Nucl. Mater.*, **10**, 2, 140 (1963).

Griesenauer N.M., Farkas M.S. and Rough F.A., "Thorium and   233
Thorium-Uranium Compounds as Potential Thermal Breeder
Fuels," US Report BMI–1680, (1964).

Griffiths L.B., "The Effect of Irradiation and Post-Irradiation Anneal- 205, 210, 211
ing on the Electrical Resistivity of Uranium Monocarbide,"
J. Nucl. Mater., **4**, 3, 336 (1961).

Griffiths L.B., "Uranium Carbide—A Summary of a Few Experiments," UK Report AERE–R4207, (1963).   205, 210, 211

Gruber E.E., "Analysis of the Effect of an Energy Gradient on Bubble Coalescence in Solids," AIME Symposium on Radiation Effects, Asheville, North Carolina, (1965).   60, 65

Hahn R.D., "A Study of Uranium Carbide and Cladding Materials for High Temperature Sodium-Cooled Reactors," US Report NAA–SR–7696, (1963).   202, 217

Hahn R.D., "Development and Evaluation of UC Alloy Fuel Materials: Irradiation Analysis—Progress Report," in US Report NAA–SR–9999, pI–B–47, (1964).   215, 223

Hanna G.L., Hickman B.S. and Hilditch R.J., "The Irradiation Behaviour of Beryllium Oxide Dispersion Fuels," Australian Report AAEC/E106, (1963).   193

Hanna G.L. and Hilditch R.J., "The Irradiation Behaviour of Cold-Pressed and Sintered Beryllium Oxide Dispersion Fuels," Australian Report AAEC/E124, (1964).   193

Harder B.R. and Sowden R.G., "The Oxidation of Uranium Dioxide by Water Vapour Under Reactor Irradiation," UK Report AERE–M725, (1960).   179

Harms W.O., "Coated-Particle Fuel Development at Oak Ridge National Laboratory," in US Report TID–7654, 71, (1962).   253

Hausner H., "Determination of the Melting Point of Uranium Dioxide," *J. Nucl. Mater.*, **15**, 3, 179 (1965).   120

Hausner H. and Nelson R.C., "Correlation of UO_2 Microstructures from In-Pile and Out-of-Pile Experiments," US Report GEAP–4535, (1965).   129

Hawkings R.C. and Bain A.S., "The Irradiation and Post-Irradiation Examination of Pressurized UO_2 Pellet Rod Mk-V," Canadian Report AECL–1790, (1963).   124

Hawkings R.C. and Robertson J.A.L., "The Thermal Conductivity of UO_2 under Irradiation—Some Early Results," Canadian Report AECL–1733, (1963).   124

Hayward B.R. and Wilkinson L.E., "Radiation Behavior of Metallic Fuels for Sodium Graphite Reactors," US Report NAA–SR–3411, (1959).   266, 267

Hesketh R.V., "A Possible Mechanism of Irradiation Creep and its Reference to Uranium," *Phil. Mag.*, Series 8, 7, 1417 (1962).   38

Hetzler F.J. and Zebroski E.L., Thermal Conductivity of Stoichiometric and Hypostoichiometric Uranium Oxide at High Temperatures," *Trans. Amer. Nucl. Soc.*, **7**, 2, 392 (1964).   128

Hicks H.G. and Armstrong J.C., "Escape of Fission Products from UO_2-Loaded BeO Fuel Elements," US Report UCRL–6148, (1961, declassified 1964).   193

Hilborn J.W., "Self-Powered Neutron Detectors for Reactor Flux Monitoring," *Nucleonics*, **22**, 2, 69 (1964).   20

Horak J.A., Kittel J.H. and Dunworth R.J., "The Effects of Irradiation on Uranium-Plutonium-Fissium Alloys," US Report ANL–6429, (1962).   276

Horak J.A., Kittel J.H. and Rhude H.V., "The Effects of Irradiation on some Binary Alloys of Thorium-Plutonium and Zirconium-Plutonium," US Report ANL–6428, (1962).   277

Horn G.R., Christensen J.A. and Pember L.A., "Identification of the Molten Zone in Irradiated UO$_2$," US Report HW–SA–3055 (Rev), (1963). Abstract in *Trans. Amer. Nucl. Soc.*, **6**, 2, 348, (1963).   117, 118

Howe J.P. and Weber C.E., "Limitations on the Performance of Nuclear Fuels," in US Report TID–2012, 163, (1957), quoted by Brinkman J.A., "Fundamentals of Fission Damage," in *Nuclear Metallurgy*—Vol. VI, p.l. AIME Publicn., (1959).   40

Howe L.M., "Irradiation Behaviour of Enriched U$_3$Si Elements Sheathed in Zircaloy-2," Canadian Report AECL–984, (1960).   231, 232

Howe L.M., Private communication, (1965).   231

Hudson B., "The Crystallography and Burgers Vectors of Dislocation Loops in α-Uranium," *Phil. Mag.*, Series 8, **10**, 949 (1964).   35

Hurst D.G., "Diffusion of Fission Gas—Calculated Diffusion from a Sphere Taking into Account Trapping and Return from the Traps," Canadian Report AECL–1550, (1962).   146

Johns I.B., "Solubility of Argon in Uranium," US Report MDDC–290, (1944).   55

Johnson M.P., Private Communication, (1962).   83

Johnson M.P. and Holland W.A., "Irradiation of U–Mo Base Alloys," US Report NAA–SR–6262, (1964).   81

Johnson W.V., "The Effects of Transients and Longer-Time Anneals on Irradiated Uranium-Zirconium Alloys," US Report KAPL–1965, (1958).   269

Jones L.J., "Radiation Stability of Uranium-Base Alloys," Part IV of "Development and Properties of Uranium-Base Alloys Corrosion Resistant in High Temperature Water," US Report WAPD–127, (1957).   99

Jones T.I., "The High Temperature Irradiation of Aluminum-Plutonium Alloys Clad in Stainless Steel," Canadian Report AECL–1467, (1962).   279

Jones T.I., "The Irradiation of Aluminum-Plutonium Alloys in Zircaloy-2 Sheathing," Canadian Report AECL–1589, (1962).   279

Karkhanavala M.D. and Carroll R.M., "In-Pile Measurement of the Electrical Resistivity and Thermo-Electric Power of Sintered UO$_2$," US Report ORNL–3093, (1961).   175

Katcoff S., "Fission Product Yields from Neutron-Induced Fission," *Nucleonics*, **18**, 11, 201 (1960).   8

Kehoe R.B., "Non-Equilibrium Beta Retention in Fuel Elements," UK Report IGR–TN/C.855, (1958).   76

Keller D.L., "Predicting Burnup of Stainless-UO$_2$ Cermet Fuels," *Nucleonics*, **19**, 6, 45 (1961).   239

Keller D.L., Cunningham G.W., Murr W.E., Fromm E.O. and Lozier D.E., "High-Temperature Irradiation Test of UO$_2$ Cermet Fuels," US Report BMI–1608, (1963).   250

Kittel J.H. and Beck W.N., "Behavior of Refractory Metal-Clad Plutonium Alloy Fuel Under Irradiation," US Report CONF–373–2, (1963).   276

Kittel J.H., Bierlein T.K., Hayward B.R. and Thurber W.C., "Irradiation Behavior of Metallic Fuels," *3rd ICPUAE*, P/239, United Nations, (1964).   44, 45, 82, 83, 85, 239, 240

Kittel J.H., Greenberg S., Paine S.H. and Draley J.E., "Effects of Irradiation on Some Corrosion-Resistant Fuel Alloys," *Nucl. Sci. Eng.*, **2**, 4, 431 (1957).   98, 99

Kittel J.H., Horak J.A., Murphy W.F. and Paine S.H., "Effects of   265, 266
Irradiation on Thorium and Thorium-Uranium Alloys," US
Report ANL–5674, (1963).

Kittel J.H., Neimark L.A., Carlander R., Kruger O.L. and Lied R.C.,   224
"Preliminary Irradiations of PuC and UC–PuC," US Report
ANL–6678, (1963).

Kittel J.H. and Paine S.H., "Effect of Irradiation on Fuel Materials,"   27, 48, 274
2nd ICPUAE, 5, 500 (1958).

Kittel J.H., Reinke C.F. and Horak J.A., "Swelling of Cast Thorium/   266
Uranium Alloys During Irradiation and Post-Irradiation Annealing," Trans. Amer. Nucl. Soc., 6, 2, 372 (1963).

Kittel J.H. and Smith K.F., "Effects of Irradiation on Some Corro-   99, 230, 232
sion-Resistant Fuel Alloys," US Report ANL–5640, (1960).

Kollie T.G., McElroy D.L., Graves R.S. and Fulkerson W., "A   128
Thermal Comparator Apparatus for Thermal Conductivity Measurements from 50 to 400°C," in Nuclear Metallurgy—Vol. X,
651, AIME Publicn., (1964).

Konobeevsky S.T., "Relaxation of Elastic Stresses Under the Ac-   37
tion of Neutron Irradiation," Atomnaya Energiya, 9, 3, 194
(1960).

Konobeevsky S.T., Dubrovin K.P., Levitsky B.M., Panteleev L.D.   37
and Pravdyuk N.F., "Some Physico-Chemical Processes Occurring
in Fissionable Materials Under Irradiation," 2nd ICPUAE, 5,
P/574, United Nations, (1958).

Konobeevsky S.T., Levitsky B.M., Panteleev L.D. and Dubrovin   78
K.P., "The Effect of Irradiation on Phase Transformations in
Cu–Sn and Cu–Sn–Pu Alloys," J. Nucl. Mater., 5, 3, 317 (1962).

Konobeevsky S.T., Pravdyuk N.F. and Kutaitsev V.I., "Effect of   26, 33, 35, 77, 79, 91,
Irradiation on the Structure and Properties of Fissionable Mater-   96
ials," 1st ICPUAE, 7, 443, United Nations, (1955).

Kramer D., Johnston W.V. and Rhodes C.G., "Reduction of Fission-   79, 87
Product Swelling in Uranium Alloys by Means of Finely Dispersed Phases," J. Inst. Metals, 93, 145 (1965).

Lacy C.E. and Leary E.A., "Irradiation Performance of Highly   268
Enriched Fuel," US Report KAPL–1952, (1958).

Lagerwall T. and Schmeling P., "Interpretation and Evaluation of   144, 149
Non-Ideal Gas Release in Post-Activation Measurements," Eur-
atom Report EUR595.e, (1964).

Lamale G.E., Hare A.W., Krause H.H., Hopkins A.K., Stang J.H.,   271
Simons E.M. and Dickerson R.F., "High Temperature Irradiation
of a Zirconium Hydride- 2w/o Uranium Alloy," US Report BMI-
1401, (1959).

Lambert J.D.B., "Irradiation Stability of Dispersion Fuels," Trans.   244
Brit. Ceram. Soc., 62, 3, 247 (1963).

Lee J.A., Ardley G.W. and Burnett R.C., Unpublished work quoted   91
by Pugh, (1955).

Leeser D.O., Rough F.A. and Bauer A.A., "Radiation Stability of   81
Fuel Elements for the Enrico Fermi Power Reactor," 2nd ICPUAE,
5, 587, United Nations, (1958).

Leggett R.D., Mastel B. and Bierlein T.K., "Irradiation Behavior   42, 44, 45
of High Purity Uranium," US Report HW–79559, (1964).

Leitz F.J., "Plutonium as a Fuel for the Fast Ceramic Reactor,"   161, 187
in US Report HW–75007, 17.1, (1962).

Levy M., Kirianenko A., Brebec G. and Adda Y., "Contribution to the Study of Rare Gas Precipitation in Metals," *Comptes Rendus*, **252**, 876 (1961). — 41

Lewis W.B., "The Return of Escaped Fission Product Gases to UO_2," Canadian Report AECL–964, (1960). — 59, 142, 148

Lewis W.B., "Designing Heavy Water Reactors for Neutron Economy and Thermal Efficiency," Canadian Report AECL–1163, (1961). — 3

Lewis W.B., "Engineering for the Fission Gas in UO_2," *Trans.. Amer. Nucl. Soc.*, **8**, 1, 24 (1965). — 150

Lewis W.B., Private communication, (1965). — 72

Lewis, W.B. MacEwan J.R., Stevens W.H. and Hart R.G., "Fission-Gas Behaviour in Uranium-Dioxide Fuel," *3rd ICPUAE*, P/19, United Nations, (1964). — 138, 144, 145, 147, 148, 151

Lindner R. and Matzke Hj., "Diffusion of Radio-Active Rare Gases in Uranium Oxide and Monocarbide," *Z. für Naturforschung*, **14a**, 12, 1074 (1959). — 144, 219, 220

Long G. Davies D. and Findlay J.R., "Diffusion of Fission Products in Uranium Dioxide and Uranium Monocarbide," in US Report TID–7610, 1, (1960). — 142

Long G., Stanaway W.P. and Davies D., "Experiments Relating to the Mechanism of the Diffusion of Xenon–133 in Uranium Dioxide," UK Report AERE–M1251, (1964). — 146

Loomis B.A., "Swelling of Aluminum-Clad Aluminum-Plutonium Alloys on Post-Irradiation Annealing," US Report ANL–6651, (1963). — 278

Loomis B.A., "Swelling of Irradiated Uranium-Zirconium Alloys on Annealing," *Nucl. Sci. Eng.*, **20**, 1, 112 (1964). — 268

Loomis B.A., Private communication, (1965). — 29

Loomis B.A., Blewitt T.H., Klank A.C. and Gerber S.B., "Elongation of Uranium Single Crystals during Neutron Irradiation," *Appl. Phys. Letters*, **5**, 7, 135 (1964). — 28

Loomis B.A. and Pracht D.W., "Swelling of Uranium on Postirradiation Annealing," *J. Nucl. Mater.*, **10**, 4, 346 (1963). — 70

Lyons M.F., Coplin D.H., Nelson R.C. and Zimmerman D.L., "UO_2 Fuel Performance—Progress Report," in US Report GEAP–3771–13, 3–1, (1964). — 20, 117, 163

Lyons M.F., Coplin D.H., Pashos T.J. and Weidenbaum B., "UO_2 Pellet Thermal Conductivity from Irradiations with Central Melting," US Report GEAP–4624, (1964). — 124, 127

MacDonald R.D., "The Effect of TiO_2 and Nb_2O_5 Additions on the Irradiation Behavior of Sintered UO_2," Canadian Report AECL–1810, (1963). — 180

MacEwan J.R., "Grain growth in Sintered Uranium Dioxide: I Equiaxed Grain Growth," *J. Amer. Ceram. Soc.*, **45**, 1, 37 (1962). — 106, 164

MacEwan J.R. and Hayashi J., "Grain Growth in UO_2: III Some Factors Influencing Equiaxed Grain Growth," British Ceramic Soc. Meeting, Harwell, England, (1965). — 108

MacEwan J.R. and Lawson V.B., "Grain Growth in Sintered Uranium Dioxide: II Columnar Grain Growth," *J. Amer. Ceram. Soc.*, **45**, 1, 42 (1962). — 109

MacEwan J.R. and Lewis W.B., Private communication, (1965), Partly Discussed by Lewis, (1965a). — 150

MacEwan J.R., Maki H., Harvey A. and Notley M.J.F., Unpublished work quoted by Robertson et al., (1964). 130

MacEwan J.R. and Stevens W.H., "Xenon Diffusion in UO_2—Some Complicating Factors," *J. Nucl. Mater.*, **11**, 1, 77 (1964). 146

Madsen P.E., "Effect of Irradiation on the Hardness of Alpha-Uranium," UK Report AERE M/R 741, (1951, declassified 1956). 91

Makin M.J., Chatwin W.H., Evans J.H., Hudson B. and Hyam E.D., "The Study of Irradiation Damage in Uranium by Electron Microscopy," in *Uranium and Graphite*, p. 45. Inst. of Metals, (1962). 34, 50, 53, 58, 87, 88, 96, 97

Markowitz J.M., Unpublished work quoted by Eichenberg J.D., Frank P.W., Kisiel T.J., Lustman B. and Vogel K.H., "Effects of Irradiation on Bulk Uranium Dioxide," US Report TID–7546, 616, (1957). 138

Matzke Hj., Verbal Presentation at Amer. Nucl. Soc. Meeting, Boston June, (1962). 219

Matzke Hj., "Diffusion in Doped UO_2," *Trans. Amer. Nucl. Soc.*, **8**, 1, 26 (1965). 142

Matzke Hj. and Lindner R., "Characteristics and Behaviour of High Temperature Nuclear Fuels," *Atomkernenergie*, **9**, 1/2, 2 (1964). 219, 220

May J.E., Notley M.J.F., Stoute R.L. and Robertson J.A.L., "Observations on the Thermal Conductivity of Uranium Oxide," Canadian Report AECL–1641, (1962). Abstract in *Trans. Amer. Nucl. Soc.*, **5**, 2, 473 (1962). 128

May J.E. and Stoute R.L., "Observations on the Thermal Conductivities of Partially-Reduced Ceramic Oxides," Canadian Report AECL–2169, (1965). 128, 180

Mehan R.L., "Bend Tests of Irradiated Fuel Element Sections—Progress Report," in US Report KAPL–2000–1, A.17, (1958). 270

Melehan J.B., Barnes R.H., Gates J.E. and Rough F.A., "Release of Fission Gases from UO_2 during and after Irradiation," US Report BMI–1623, (1963). 138, 139, 141

Melehan J.B. and Gates J.E., "In-Pile Fission-Gas Release from Uranium Carbide and Uranium Nitride," US Report BMI–1701, (1964). 219, 220, 228

Mitelman M.G., Erofeev R.S. and Rozenblyum N.D., "Transformation of the Energy of Short-Lived Isotopes," *Atomnaya Energiya*, **10**, 72 (1961). 20

Mogard H., Djurle S., Multer I., Myers H.P., Nelson B. and Runfors U., "Fuel Development for Swedish Heavy Water Reactors," *3rd ICPUAE*, P/608, United Nations, (1964). 122

Morgan C.S., Fitts R.B. and Scott J.L., "Influence of Stress on Fission Gas Release," *J. Amer. Ceram. Soc.*, **48**, 3, 166 (1965). 146

Morgan W.W., Hart R.G., Jones R.W. and Edwards W.J., "A Preliminary Report on Radial Distribution of Fission-Product Xenon and Cesium in UO_2 Fuel Elements," Canadian Report AECL–1249, (1961). 186

Mustelier J.-P., "Some Results From Irradiations of Fuels Considered for Rapsodie," in *Radiation Damage in Reactor Materials*, p. 163. IAEA, Vienna, (1962). 275

Mustelier J.-P., Mikailoff H., Ratier J.L., Gondal D. and Bloch J., "The Effects of Irradiation on Uranium Alloys with Small Molybdenum Contents," in *Properties of Reactor Materials and the Effects of Radiation Damage*, p. 505. Butterworths, London, (1961). 83

Naymark S. and Spalaris C.N., "Oxide Fuel Fabrication and Perfor-   159
mance," *3rd ICPUAE*, P/233, United Nations, (1964).

Neimark L.A., "Recrystallization in ThO$_2$-UO$_2$ During Irradiation,"   181
Trans. Amer. Nucl. Soc., **5**, 1, 226 (1962).

Neimark L.A. and Carlander R., "Irradiation of PuC and UC–20   225
w/o PuC—Progress Report," in US Report ANL–6868, 53, (1964).

Neimark L.A. and Carlander R., "Irradiation Effects in Uranium   228
Sulfide," in *Nuclear Metallurgy*—Vol. X, 753, AIME Publicn.,
(1964).

Neimark L.A., Kittel J.H. and Hoenig C.L., "Irradiation of Metal-   183
Fiber-Reinforced Thoria-Urania," US Report ANL–6397, (1961).

Neimark L.A. and Kittel J.H., "The Irradiation of Aluminum Alloy-   181, 182
Clad Thoria-Urania Pellets," US Report ANL–6538, (1964).

Nelson R.S., "The Observation of Collision Sequences in UO$_2$,"   7
J. Nucl. Mater., **10**, 2, 154, (1963).

Newkirk H.W., Daniel J.L. and Mastel B., "Electron Microscope   113
Studies of Damage in Irradiated Uranium Dioxide," *J. Nucl.
Mater.*, **2**, 3, 269 (1960)

Nilsson G., "Ejection of Uranium Atoms from UO$_2$ by Fission   140
Fragments," Swedish Report AE–136, (1964).

Notley M.J.F., "The Relative Axial Expansions under Irradiation of   134
Stacks of UO$_2$ Pellets in Zircaloy Sheaths," Canadian Report
AECL–1598, (1962).

Notley M.J.F., "The Thermal Conductivity of Columnar Grains in   129
Irradiated UO$_2$ Fuel Elements," Canadian Report AECL–1822,
(1963).

Notley M.J.F., Bain A.S. and Robertson J.A.L., "The Longitudinal   134, 135
and Diametral Expansions of UO$_2$ Fuel Elements," Canadian
Report AECL–2143, (1964).

Notley M.J.F. and Fitzsimmons W.D.C., "The Irradiation of Hy-   133
draulic Rabbit Specimens to Study Sheath Deformations," Canadian
Report AECL–1664, (1962).

Notley M.J.F. and Harvey A., "The Length Changes of Zircaloy   20, 133, 134
Sheathed Fuel Elements during Irradiation," Canadian Report
AECL–1846, (1963).

Notley M.J.F. and MacEwan J.R., "The Effect of UO$_2$ Density on   109, 130, 137
Fission Product Gas Release and Sheath Expansion," Canadian
Report AECL–2230, (1965).

Osthagen K.H. and Bauer A.A., "Chemical Diffusion in UC$_2$ and   202
UC and Thermal Diffusion in UC," US Report BMI–1686, (1964).

Padden T.R., Unpublished work quoted by Belle, (1961).   170

Padden T.R. and Schnizler P.S., "Microstructural Changes Asso-   114, 165, 170
ciated with Irradiation Burnup of Uranium Dioxide," in *5th
International Congress for Electron Microscopy*, p. G–14, Academic
Press, New York, (1962): Full text as US Report WAPD–T–1501.

Paine S.H. and Kittel J.H., "Irradiation Effects in Uranium and its   26, 41, 91, 96
Alloys," *1st ICPUAE*, 7, 445, United Nations, (1955).

Papathanassopoulos K., Thesis on Diffusion of Fission Xenon in   220
Uranium Carbide, Technical University of Braunschweig, West
Germany, (1963).

Paprocki S.J., Dickerson R.F., Cunningham G.W., Murr W.E. and   224
Lozier D.E., "Fabrication and Irradiation of SM–2 Core Materials,"
US Report BMI–1528, (1961).

Pardue W.M., Storhok V.W., Smith R.A. and Keller D.L., "An Evaluation of Plutonium Compounds as Nuclear Fuels," US Report BMI–1698, (1964). — 233

Parker G.W., Creek G.E., Martin W.J. and Barton C.J., "Fuel Element Catastrophe Studies: Hazards on Fission Product Release from Irradiated Uranium," US Report ORNL–CF–60–6–24, (1960). — 100

Pashos T.J., de Halas D.R., Keller D.L. and Neimark L.A., "Irradiation Behavior of Ceramic Fuels," 3rd ICPUAE, P/240, United Nations, (1964). — 128, 178, 192, 229, 233

Powers R.M., Cavallaro Y. and Mathern J.P., "The Effect of Solid Solution Additions on the Thermal Conductivity of UO_2," US Report SCNC–317, (1960). — 180

Pugh S.F., "Damage Occurring in Uranium During Burnup," 1st ICPUAE, 7, 441, United Nations, (1955). — 26, 41

Pugh S.F., "Swelling in Alpha Uranium Due to Irradiation," J. Nucl. Mater., 4, 2, 177 (1961). — 41, 47, 48

Pugh S.F., "Swelling in Alpha-Uranium Due to Irradiation in the Range 400 to 650°C," J. Nucl. Mater., 12, 3, 355 (1964). — 29

Pugh S.F. and Butcher B.R., "Metallic Fuels," in US Report TID–8540, 331, (1964). — 34, 84, 85, 100

Quéré Y., "Effects of Light Irradiations on Uranium," J. de Physique, 24, 489, (1963). — 91

Quéré Y., "Resistivity Changes in Uranium Irradiated at Low Temperatures," J. Nucl. Mater., 9, 3, 290 (1963). — 91

Quéré Y. and Doulat J., "Deformation of Uranium Under Irradiation at Low Temperature," Comptes Rendus, 252, 1305 (1961). — 28

Quéré Y. and Nakache F., "Evaluation of the Volume of a Fission Spike in Uranium," J. Nucl. Mater., 1, 2, 203 (1959). — 91

Rabin S.A. and Ullmann J.W., "High Burnup ThO_2–UO_2 Pellet Rods—Progress Report," US Report ORNL–3670, 242, (1964). — 183

Rao S.V.K., "Investigation of ThO_2–UO_2 as a Nuclear Fuel," Canadian Report AECL–1785, (1963). — 181, 182

Reinke C.F., "Irradiation and Post-Irradiation Annealing of some Aluminum-Base Fuels," US Report ANL–6665, (1963). — 248, 249, 261, 262

Reinke C.F. and Carlander R., "Examination of Irradiated EBWR Core-I Fuel Elements," US Report ANL–6091, (1960). — 85

Reynolds M.B., "Fission Gas Behavior in the Uranium-Aluminum System," Nucl. Sci. Eng., 3, 4, 428 (1958). — 262

Reynolds M.B., "The Measurement of Free Fission Gas Pressure in Operating Reactor Fuel Elements," US Report GEAP–4135, (1963). — 20, 151

Rich J.B. and Barnes R.S., "Gas Bubble Growth on Heating Irradiated Uranium," UK Report AERE–R2935, (1959). — 67

Richt A.E. and Schaffer L.D., "Army Reactors Program Annual Progress Report for Period Ending October 31, 1962," US Report ORNL–3386, 53, (1963). — 241

Rimmer D.E. and Cottrell A.H., "The Solution of Inert Gas Atoms in Metals," Phil. Mag., Series 8, 2, 1345 (1957). — 54

Roake W.E., "Irradiation Alteration of Uranium Dioxide," in Radiation Damage in Reactor Materials, p. 429. IAEA, Vienna, (1962). — 164, 174, 178

Roberts A.C. and Cottrell A.H., "Creep of α-Uranium during Irradiations with Neutrons," Phil. Mag., Series 8 1, 711 (1956). — 36

Roberts L.E.J., Brock P., Findlay J.R., Frost B.R.T., Russell L.E., Sayers J.B. and Wait E., "The Behaviour of UO_2 and of (U, Pu)O_2 Fuel Materials Under Irradiation," *3rd ICPUEA*, P/155, United Nations, (1964). 164, 167

Robertson J.A.L., "$\int k.d\theta$ in Fuel Irradiations," Canadian Report AECL–807, (Revised 1961). 19

Robertson J.A.L., "Interesting developments in UO_2 Technology," in *New Nuclear Materials Including Non-Metallic Fuels*, Vol. II p. 57, IAEA, Vienna, (1963). 138

Robertson J.A.L., Bain A.S., Booth A.H., Howieson J., Morison W.G. and Robertson R.F.S., "Behaviour of Uranium Oxide as a Reactor Fuel," *2nd ICPUAE*, **6**, 655, United Nations, (1958). 150, 176, 179

Robertson J.A.L., Bain A.S. and Ridal A., "The Effect of 4mol% Y_2O_3 on the Thermal Conductivity of UO_2," Canadian Report AECL–1037, (1960). 180

Robertson J.A.L., Ross A.M., Notley M.J.F. and MacEwan J.R., "Temperature Distribution in UO_2 Fuel Elements," *J. Nucl. Mater.*, **7**, 3, 225 (1962). 103, 106, 121, 124, 125, 127, 132, 133

Robertson R.F.S., "Tests of Defected Thoria-Urania Fuel Specimens in EBWR," US Report ANL–6022, (1960). 183

Rogers M.D., "Dynamic Equilibrium Between Ejection and Reejection of Uranium by Fission Fragments," *J. Nucl. Mater.*, **12**, 3, 332 (1964). 139

Rogers M.D., "Mass Transport of Uranium by Fission Fragments," *J. Nucl. Mater.*, **15**, 1, 65 (1965). 139

Rogers M.D., "Mass Transport and Grain Growth Induced by Fission Fragments in Thin Films of Uranium Dioxide," *J. Nucl. Mater.*, **16**, 3, 298 (1965). 140

Rogers M.D. and Adam J., "Ejection of Atoms from Uranium by Fission Fragments," *J. Nucl. Mater.*, **6**, 2, 182 (1962). 139

Rogers M.D. and Adam J., Private communication, (1962). 227

Rose H.C., "A Compressive Creep Test of Alpha-Uranium under Neutron Irradiation," *J. Inst. Metals*, **86**, 122 (1957). 39

Ross A.M., "The Dependence of the Thermal Conductivity of Uranium Dioxide on Density, Microstructure, Stoichiometry and Thermal Neutron Irradiation," Canadian Report AECL–1096, (1960). 122, 123

Rothwell E., "The Release of Kr^{85} from Irradiated Uranium Dioxide on Post-Irradiation Annealing," *J. Nucl. Mater.*, **5**, 2, 241 (1962). 144

Rough F.A. and Chubb W., "An Evaluation of Data on Nuclear Carbides," US Report BMI–1441, (1960). 214, 217

Rufeh F., "Solubility of Helium in Uranium Dioxide," US Report UCRL–11043, (1964). 147

Runnalls O.J.C., "Irradiation Experience with Rods of Plutonium-Aluminum Alloy," *2nd ICPUAE*, **6**, 475, United Nations, (1958). 278

Runnalls O.J.C., "Plutonium Fuel Program in Canada," in US Report HW–75007, 8.1, (1962). 187

Runnalls O.J.C., Private communication, (1965). 226

Runnalls O.J.C. and Boucher R., "Transformations in UAl_4 and $PuAl_4$," *Trans. Met. Soc. AIME*, **233**, 9, 1726 (1965). 264

Savage J.W., "Diffusion of Fission Gas in Uranium," US Report NAA–SR–6761, (1963). 56

Sayers J.B., Rose K.S.B., Coobs J.H., Hauser G.P. and Vivante C.,   253, 254, 256, 258, 259
"The Irradiation Behaviour of Coated Particle Fuel," in *Carbides in Nuclear Energy*, Vol. 2, p. 919, Macmillan and Co. Ltd., London, (1963).

Sayers J.B. and Worth J.H., "Comparison of the Irradiation Behaviour   184
of 1 % PuO_2 in UO_2 and Stoichiometric UO_2," in *Power Reactor Experiments*, Vol. 1, p. 171, IAEA, Vienna, (1961).

Schurenkamper A. and Soulhier R., "New Results on the Behaviour   139
of Fission Gases at High Temperature in UO_2 under Irradiation," French Report CEA–R2588, (1964).

Seeger A., "The Nature of Radiation Damage in Metals," in *Radiation*   6
Damage in Solids, Vol. 1, p. 101, IAEA, Vienna, (1962).

Shaked H., Olander D.R. and Pigford T.H., "Diffusion of Xe–133 in   220
Uranium Monocarbide," *Trans. Amer. Nucl. Soc.*, **6**, 1, 131 (1963).

Shaw D., "The Tensile Properties of Irradiated Uranium," *Nucl.*   92
Eng., **5**, 214 (1960).

Shewmon P.G., "The Movement of Small Inclusions in Solids by a   60
Temperature Gradient," *Trans. Met. Soc. AIME*, **230**, 5, 1134 (1964).

Shiriaeva L.V. and Tolmachev Iu.M., "The Chemical Behaviour of   166
Mo–99 formed by Irradiation of Uranium Compounds by Neutrons," *Atomnaya Energiya*, **3**, 318 (1957): "The Chemical Behaviour of Mo–99 formed by the Slow Neutron Fission in Uranium Oxides," ibid., **6**, 528 (1959): Also, *J. Nucl. Energy* A, **12**, 181 (1960).

Shoudy A.A., McHugh W.E. and Silliman M.A., "The Effect of   82
Irradiation Temperature and Fission Rate on the Radiation Stability of Uranium-10 wt% Molybdenum Alloy," in *Radiation Damage in Reactor Materials*, p. 133, IAEA, Vienna, (1962).

Simnad M.T., "Irradiation Effects on Dispersion Type BeO–UO_2 192
Fuel for EBOR," US Report GA–4643 Part II, (1963).

Simnad M.T., Hopkins G. and Shoptaugh J., "Fuel Elements for the   272
Advanced TRIGA-Mark III Pulsing Reactor," *Trans. Amer. Nucl. Soc.*, **7**, 1, 110 (1964).

Sinizer D.I., Webb B.A. and Berger S., "Irradiation Behavior of   202, 215
Uranium Carbide Fuels," in *Radiation Damage in Reactor Materials*, p. 287, IAEA, Vienna, (1962).

Smith K.F., "Irradiation of Uranium-Fissium Alloys and Related   83
Compositions," US Report ANL–5736, (1957).

Speight M.V., "Bubble Diffusion and Coalescence during the Heat   60
Treatment of Materials Containing Irradiation-Induced Gases," *J. Nucl. Mater.*, **12**, 2, 216 (1964).

Speight M.V., "The Migration of Gas Bubbles in Material Subject   60
to a Temperature Gradient," *J. Nucl. Mater.*, **13**, 2, 207 (1964).

Speight M.V. and Greenwood G.W., "Grain Boundary Mobility   71
and its Effects in Materials Containing Inert Gases," *Phil. Mag.*, **9**, 100, 683 (1964).

Stahl D. and Strasser A., "Properties of Solid Solution Uranium-   225
Plutonium Carbides," in *Carbides in Nuclear Energy*, Vol. 1, p. 371, Macmillan and Co. Ltd., London, (1963).

Stevens W.H., Unpublished work quoted by Frost (1963).   221

Stevens W.H., MacEwan J.R. and Ross A.M., "The Diffusion   141, 143, 144
Behaviour of Fission Xenon in Uranium Dioxide," in US Report TID–7610, 7, (1960).

Stewart J.C.C., Eldred V.W., Heal T.J., Pickman D.O. and Rogan H.,   46
"Review of Progress of Development, Manufacture and Performance of Magnox Fuel Elements in the United Kingdom," *3rd ICPUAE*, P/560, United Nations, (1964).

Stora J.-P., deB. de Sigoyer B., Delmas R., Deschamps P., Lavaud B.   125
and Ringot C., "Thermal Conductivity of Sintered Uranium Oxide Under In-Reactor Conditions," French Report CEA–R2586,(1964).

Storhok V.W., "Fabricating Plutonium for Better Performance,"   90
Nucleonics, **21**, 1, 38 (1963).

Strasser A., Cihi J., Sheridan W. and Storhok V., "Irradiation   225
Behavior of Solid-Solution Uranium-Plutonium Monocarbides," in *Nuclear Metallurgy*—Vol. X, p. 729, AIME Publicn., (1964).

Sykes E.C. and Greenwood G.W., "Tensile Creep Ductility of Neu-   92
tron Irradiated Alpha-Uranium," *Nature*, **206**, 4980, 181 (1965).

Tardivon D., "Crystallographic Study of Radiation-Induced Recovery   96
in Cold-Worked Uranium," in *Nuclear Metallurgy*—Vol. VI, 39, AIME Publin., (1959).

Tennery V.J., "Review of Thermal Conductivity and Heat Transfer   180
in UO_2," US Report ORNL–2656, (1959).

Thomas D.E., Fillnow R.H., Goldman K.M., Hino J., Van Thyne   85
R.J., Holtz F.C. and MacPherson D.J., "Properties of Gamma-Phase Alloys of Uranium," *2nd ICPUAE*, **5**, 610, United Nations, (1958).

Thompson M.W., "An Experiment to Clarify the Role of Point   33
Defects in the Radiation Growth of α-Uranium," *J. Nucl. Mater.*, **3**, 3, 354 (1961).

Townley C.W., Miller N.E., Ritzman R.L. and Burian R.J., "In–Pile   252, 255
Performance of Ceramic Coated-Particle Fuels," *Nucleonics*, **22**, 2, 43 (1964).

Troutner V.H., "Mechanism and Kinetics of Uranium Corrosion and   99
Uranium Core Fuel Element Ruptures in Water and Steam," US Report HW–67370, (1960).

Tucker C.W. and Senio P., "On the Nature of Thermal Spikes,"   75
J. Appl. Phys., **27**, 3, 207 (1956).

Vetrano J.B., "A Uranium Dispersion, Zirconium Hydride Matrix Fuel   271
for Space Power Systems," *Trans. Amer. Nucl. Soc.*, 3, 2, 444, (1960).

Waldron M.B., Adwick A.G., Lloyd H., Notley M.J.F., Poole D.M.,   274, 277, 278
Russell L.E. and Sayers J.B., "Plutonium Technology for Reactor Systems," *2nd ICPUAE*, **6**, 690, United Nations, (1958).

Wallace W.P., Simnad M.T. and Turovlin B., "Fabrication and   271
Properties of Uranium-Zirconium Hydride Fuel Elements for TRIGA Reactors," in US Report TID–7559 Part I, 70, (1958).

Weber C.E., "Progress on Dispersion Elements," in *Progress in*   238, 239
Nuclear Energy—Series V, Vol. 2, p. 295 (1959).

Weil L. and Cohen J., "Paramagnetic Susceptibility of Irradiated   175
UO_2," *J. de Physique*, **24**, 1, 76 (1963).

Weinberg C. and Quéré Y., "Influence of Cold-Work-Induced De-   29
fects on the Irradiation Growth of α-Uranium," in *VIeme Colloque de Métallurgie*, p. 35, CEN Saclay, (1962).

Westcott C.H., "Effective Cross-Section Values for Well-Moderated   11
Thermal Reactor Spectra, Third Edition," Canadian Report AECL–1101, (1960).

Whapham A.D., "Electron Microscope Observations of the Fission-   116

Gas Bubble Distribution in UO₂," *Trans. Amer. Nucl. Soc.*, **8**, 1, 21 (1965).

Whapham A.D. and Makin M.J., "The Nature of Fission Fragment Tracks in Uranium Dioxide," *Phil. Mag.* Series 8, **7**, 1441, (1962).   114

Whapham A.D. and Sheldon B.E., "Transmission Electron Microscope Study of Irradiation Effects in Sintered Uranium Dioxide," *J. Nucl. Mater.*, **10**, 2, 157 (1963).   116

Williams C.D., Unpublished work quoted by MacDonald R.D. in Presenting paper at Amer. Nucl. Soc. Meeting, San Francisco, Nov.-Dec., (1964).   164

Williams J., "Dispersions of Oxides in Oxide Matrices as High-Temperature Reactor Fuels," in *New Nuclear Materials Including Non-Metallic Fuels*, Vol. II, p. 79. IAEA, Vienna, (1963).   189, 192

Williamson G.K. and Cornell R.M., "The Behaviour of Fission Product Gases in Uranium Dioxide," UK Report CEGB.RD/B/N.279, (1964).   147

Williamson H.E. and Hoffman J.P., "A Uranium Dioxide Fuel Rod Center Melting Test in the Vallecitos Boiling Water Reactor," US Report GEAP-4408, (1963).   163

Willis A.H., "The Irradiation Stability of Low wt. pct Uranium-Zirconium Alloys," *Trans. Met. Soc. AIME*, **215**, 245 (1959).   268

Willis B.T.M., "Positions of the Oxygen Atoms in UO₂.₁₃," *Nature*, **197**, 4869, 755 (1963).   101

Willis B.T.M., "Neutron Diffraction Studies of the Actinide Oxides. II Thermal Motions of the Atoms in Uranium Dioxide and Thorium Dioxide Between Room Temperature and 1100°C," *Proc. Roy. Soc.*, **274 A**, 1356, 134 (1963).   101

Winegard W.C., "Fundamentals of the Solidification of Metals," *Met. Reviews*, **6**, 21, 57 (1961).   110

Wittells M.C. and Sherrill F.A., "Fission Fragment Damage in Zirconia," *Phys. Rev. Letters*, **3**, 176 (1959).   196

Wright T.R., Kizer D.E. and Keller D.L., "Studies in the UO₂–ZrO₂ System," US Report BMI–1689, (1964).   194

Zaimovsky A.S., Sergeev G.Y., Titova V.V., Levitsky B.M. and Sokurksky Y.N., "Influence of the Structure and Properties of Uranium on its Behavior Under Irradiation," *2nd ICPUAE*, **5**, 566, United Nations, (1958).   20, 33, 36, 37, 80, 92

Additional General Sources of Information on Nuclear Fuels

Reactor Handbook, *Second Edition*, *Vol. I—Materials*, Editor, Tipton C.R., Interscience, New York, (1960).

Melehan J.B., Storhok V.W., Burian R.J., Kangilaski M. and Wullaert R.A., "The Effect of Nuclear Radiation on Ceramic Reactor-Fuel Materials," US Report REIC R.27, (1963).

Bauer A.A., Farkas M.S. and Storhok V.W., "The Effect of Nuclear Radiation on Metallic Fuel Materials," US Report REIC R.29, (1963).

Uhlmann W., "Radiation Effects on Solid, Uranium Based Nuclear Fuel Materials—a Bibliography—Vols. 1 to 4," Swedish Report VDIT–16, (1961–1964).

Reactor Materials, Prepared by Dayton R.W., Simons E.M. and Associates for the USAEC Division of Technical Information, (Quarterly).

Index

(For Author Index see References)

A

Activity profile, UO_2, 163
"Adjusted" uranium, 85
Aligned pores, uranium, 47
Al_2O_3–UO_2, 187
Al–Pu, 278
Al–U, 261
Anomalous intercept, 144
Autoradiography, 162, 165, 187

B

BeO–UO_2, 190
BeO–$(U,Th)O_2$, 190
Blistering, cermets, 240, 246, 248
  U–Al, 262
Boiling, UO_2, 110
Breakaway swelling, UO_2, 153, 158
  uranium, 44, 72, 86
Bubbles, see Pores
Bulges, cermets, 239, 240, 246, 248
  U–Al, 262
Burnup, 9
Burnup measurement, 19
Burnup units, conversion table, 14, 15
Burst of gas, 144

C

Carbide fuels, 200
Cermets, 238, 245, 248
Chanelling, 6
Circumferential ridges, 137
Coated particles, 251
Columnar grains, MgO–PuO_2, 199
  ThO_2–UO_2, 181, 184
  UO_2, 108, 129, 176, 186
  uranium, 96

Compacted powder, 175, 225
Composite diffusion coefficient, 143
Conversion table, burnup units, 14, 15
Corrosion, UO_2, 179
  uranium, 98
  U_3Si, 232
  Zr–U, 270
Cracks, cermets, 239, 243, 246
  UC, 201, 217, 218
  UO_2, 103, 125, 131, 137, 149, 241
  uranium, 44, 71, 93, 96
Creep, UO_2, 133, 151, 242
  uranium, 25, 35
  ZrO_2–UO_2, 197
Crystal structure, Al_2O_3, 189
  UC, 200, 208, 212, 213
  UC_2, 201
  U_2C_3, 201, 213
  UO_2, 101, 166
  U_3O_8, 173
  uranium, 22, 98
  ZrO_2, 196

D

Damage mechanisms, 4, 114, 173, 205
Deformation and gas release, 146
Density, UC, 209, 212, 215, 253
  UO_2, 122, 130, 137, 143, 148, 152
  UO_2–PuO_2, 185
  uranium, 23, 25, 40
  U_3Si, 229
  ZrO_2–UO_2, 195
Diametral expansion, UO_2, 134
Diffusion, irradiation–enhanced, 78
Diffusion mechanisms, UO_2, 142, 146
Diffusion near surfaces, 140, 220
Diffusion, carbon in UC, 202
  fission products in UO_2, 138, 161

305